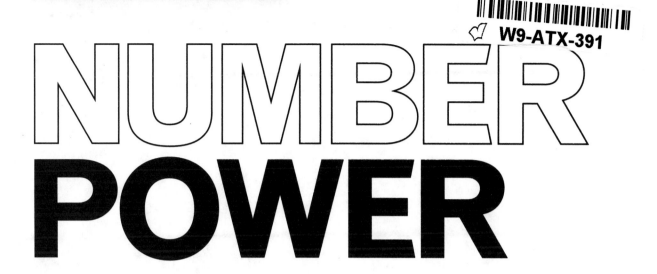

W9-ATX-391

NUMBER POWER

A REAL WORLD APPROACH TO MATH

Algebra

ROBERT MITCHELL

 **Education**

Bothell, WA • Chicago, IL • Columbus, OH • New York, NY

www.mheonline.com

 **Education**

Send all inquiries to:
Contemporary/McGraw-Hill
130 E. Randolph, Suite 400
Chicago, IL 60601

ISBN: 978-0-07-659228-9
MHID: 0-07-659228-6

Printed in the United States of America.

1 2 3 4 5 6 7 8 9 10 RHR 15 14 13 12 11

The *McGraw·Hill* Companies

TABLE OF CONTENTS

One-Step Equations

Multistep Equations

Special Equations

Factoring

Algebra Posttest A

Using Number Power

TO THE STUDENT

Welcome to *Algebra*:

Algebra is a powerful tool that makes solving complicated problems much easier. Algebra is the basic mathematical language of technical fields from auto mechanics to nursing and electronics. Also, algebra is a standard section on almost all educational and vocational tests, including GED, college entrance, civil service, and military entrance.

Number Power Algebra is designed to prepare you for taking a test and for pursuing further education or training. The book begins with a pretest that will give you an overview of the concepts you will study in this book. The first part of the book, Building Number Power, provides step-by-step instruction in the fundamentals of algebra. This part is divided into 10 chapters. Each chapter ends with a review that you can use to check your understanding. At the end of Building Number Power is Posttest A, which asks you to solve problems and show your answers.

The second part of the book, Using Number Power, will give you a chance to apply algebraic skills in real-life situations. These applications are fun and will give you added chances to see the power of algebra. At the end of Using Number Power is Posttest B, which uses a multiple-choice format much like many educational and vocational tests you may take later.

Learning how and when to use a calculator is an important skill to develop. In real life, problem solving often involves the smart use of a calculator—especially when working with large numbers. You need to know when an answer on a calculator makes sense, so be sure to have an estimated answer in mind. Remember that a calculator can help you only if you set up the problem and use the calculator correctly.

To get the most from your work, do each problem carefully. Check your answers with the answer key at the back of the book to make sure you are working accurately. Inside the back cover is a chart to help you keep track of your score on each exercise.

We hope you enjoy *Number Power Algebra!*

Algebra Pretest

This pretest will tell you which chapters of *Number Power Algebra* you need to work on and which you already have mastered. Do all the problems that you can. There is no time limit. Check your answers with the answer key at the back of the book. Then use the chart at the end of the test to find the pages where you need to work.

Write the addition shown by the number arrows.

1. _____

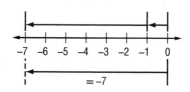

2. _____

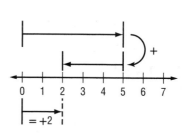

Subtract.

3. $8 - (-3) =$

4. $-12 - (-6) =$

Multiply or divide as indicated.

5. $(-6)(5) =$

6. $(-14) \div (-7) =$

Find the value of each power.

7. $4^2 =$

8. $(-3)^3 =$

Write each product or quotient as a single base and exponent.

9. $(6^4)(6^2) =$

10. $\dfrac{7^5}{7^2} =$

Find each square root below.

11. $\sqrt{64} =$

12. $\sqrt{\dfrac{25}{49}} =$

13. Write the expression "3 times the quantity n plus 6" as an algebraic expression.

Find the value of each algebraic expression.

14. $8x - 4$ when $x = 3$

15. $(a + 2)(b - 2)$ when $a = 6, b = 4$

16. Use the formula $°C = \dfrac{5}{9}(°F - 32)$ to find the Celsius temperature ($°C$) when the Fahrenheit temperature ($°F$) is $212°F$.

Circle the equation that is _not_ represented by the drawing.

17.

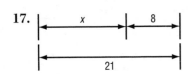

 a. $21 - x = 8$
 b. $x + 8 = 21$
 c. $x - 8 = 21$
 d. $21 - 8 = x$

For problems 18–22, solve each equation. Show all steps.

18. **a.** $x + 9 = 26$ **b.** $y - 5 = 17$

19. **a.** $3n = 18$ **b.** $\dfrac{n}{4} = 9$

20. **a.** $3x - 5 = 22$ **b.** $\dfrac{s}{4} + 8 = 20$

21. **a.** $5y + 3y = 64$ **b.** $4n - 7 = 2n + 9$

22. **a.** $2(x - 3) = 20$ **b.** $4(b + 2) = 2(b + 9)$

23. The ratio of two numbers is 4 to 3. The sum of the numbers is 49. What are the two numbers?

24. A recipe for 6 servings of dessert calls for 2 cups of flour. Using a proportion, determine how much flour is needed to make 16 servings.

25. Find the two correct values of x in the equation $3x^2 = y$ when $y = 75$.

26. Graph the line containing the points $(0, -4)$, $(2, 0)$, and $(3, 2)$.

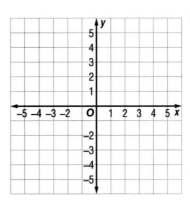

27. What is the slope of line M?

$$\text{slope} = \frac{\text{change in } y}{\text{change in } x}$$

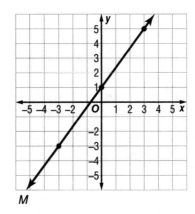

M

28. Graph the equation $y = 2x - 1$.
As a first step, complete the Table of Values.

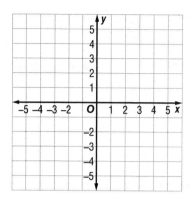

Table of Values	
x	**y**
0	
1	
2	

Write the inequality graphed on the number line.

29.

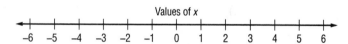

Values of *n*

−6 −5 −4 −3 −2 −1 0 1 2 3 4 5 6

Solve the following inequalities.

30. $n + 4 \geq 6$

31. $3x - 5 < 16$

32. Graph the inequality $-3 \leq x < 5$.

Values of *x*

−6 −5 −4 −3 −2 −1 0 1 2 3 4 5 6

33. Circle the pair of like terms in the group of five terms below.

$-3x^2$ $3y^2$ $3x$ $5x^2$ $-2y^3$

Add or subtract as indicated.

34. $(5x^2 + 3x) + (-2x^2 - 4x)$

35. $(3a^2 - 2b) - (2a^2 + 3b)$

Multiply or divide as indicated.

36. $2n(3n^2 + 4n - 2)$

37. $\dfrac{(14x^4 + 6x^3)}{2x^2}$

Simplify each square root.

38. $\sqrt{75} =$

39. $\sqrt{36x^4y} =$

Factor each expression as completely as possible.

40. $4x^2 + 10$

41. $12z^3 + 16z^2$

Simplify each quotient by dividing out the common factor.

42. $\dfrac{9r^2 + 12}{3} =$

43. $\dfrac{7c - 7d}{c - d} =$

ALGEBRA PRETEST CHART

This pretest can be used to identify algebra skills in which you are already proficient. If you miss only 1 question in a chapter, you may not need further study in that chapter. However, before you skip any lessons, complete the review at the end of that chapter. The end-of-chapter review will be a more precise indicator of your skill level.

<u>Problem Numbers</u>	<u>Skill Area</u>	<u>Practice Pages</u>
1, 2, 3, 4	adding and subtracting signed numbers	12–17
5, 6	multiplying and dividing signed numbers	18–21
7, 8	evaluating a power	26–28
9, 10	simplifying a product or quotient	30–31
11, 12	finding a square root	34–35
13	writing algebraic expressions	42–43
14, 15	evaluating algebraic expressions	46–48
16	evaluating formulas	49–50
17	picturing an equation	56–57
18	solving addition and subtraction equations	59–61
19	solving multiplication and division equations	63–64
20	solving multistep equations	72–75
21	solving an equation with separated variables	76–77
22	solving an equation with parentheses	82–83
23, 24	using ratios and proportions	92, 94–95
25	solving a quadratic equation	102–103
26	graphing a line	111
27	finding the slope of a line	112–113
28	graphing a linear equation	114–115
29, 32	graphing an inequality	125–126, 131–132
30, 31	solving an inequality	127–128
33	like terms in a polynomial	137
34, 35	adding and subtracting polynomials	138–141
36, 37	multiplying and dividing polynomials	142–147
38, 39	simplifying a square root	155
40, 41, 42, 43	factoring and dividing	156–159

BUILDING
NUMBER
POWER

Introducing Signed Numbers

The study of algebra begins with **signed numbers**—the set of all positive and negative numbers. Look at the thermometer at the right. Both positive and negative numbers are shown.

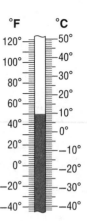

- **Positive numbers** are greater than zero and are preceded by a positive sign (+) or with no sign at all. Positive temperatures are greater than zero degrees.
- **Negative numbers** are less than zero and are always preceded by a negative sign (−). Negative temperatures are less than zero degrees.

The numbers you may be most familiar with are positive numbers. Negative numbers, though not as common as positive numbers, also have many uses.

- 15° below 0° is written as −15°.
- A $\frac{1}{2}$ dollar drop in a stock market price is written $-\frac{1}{2}$.
- 35 feet below sea level is written as −35 feet.

Zero (0) separates positive numbers from negative numbers. Zero is neither positive nor negative.

Positive 20 is written as +20 or 20.   Negative 20 is written as −20.

Positive $5\frac{3}{4}$ is written as $+5\frac{3}{4}$ or $5\frac{3}{4}$.   Negative 7.5 is written as −7.5.

Zero has no sign and is always written as 0.

(**Remember:** A non-zero number without a sign is always a positive number.)

Write each number as a signed number with a + sign or a − sign.

1. positive 8 | positive 3 | positive 12 | positive 7

2. negative 5 | negative 4 | negative 19 | negative 8

3. positive $2\frac{1}{2}$ | positive 4.75 | negative $5\frac{2}{3}$ | negative $\frac{1}{4}$

Identify each number as either positive or negative.

4. +8 | −7 | $\frac{3}{4}$ | −6

5. $2\frac{2}{3}$ | $+4\frac{1}{2}$ | −9.3 | $-\frac{3}{8}$

The Number Line

Signed numbers can be pictured on a **number line.** The value of numbers increases as you move from left to right.

Positive numbers are written to the right of 0.
- Every positive number is greater than 0.
- Every positive number is greater than any negative number.

Negative numbers are written to the left of 0.
- Every negative number is less than 0.
- Every negative number is less than any positive number.
- You can read a negative number in any of three ways.

> **Did You Know ...?**
> Whole numbers, fractions, and decimals can all be represented on a number line.

EXAMPLE   −4 can be read as *negative 4, minus 4,* or *4 below zero*.

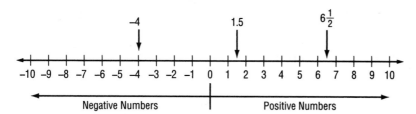

A **number arrow** can represent a signed number.
- A positive number arrow points to the right.
- A negative number arrow points to the left.

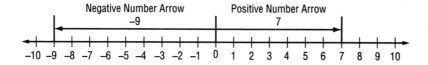

Locate these points on the number line. $+5, -4, \frac{1}{2}, 2, -3\frac{1}{4}, 6.5, -7.75, +8, -2$

1.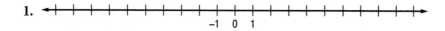

Draw a number arrow for each signed number.

2. Positive 3

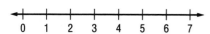

3. Negative 4

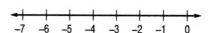

Adding Signed Numbers

In arithmetic, you learn to add and subtract positive numbers. In algebra, you learn that you can also add and subtract negative numbers. To avoid confusion, a signed number may be enclosed in parentheses. Look at these examples.

Sum or Difference	Meaning	Simplified Way of Writing
+5 + (+6)	positive 5 plus positive 6	5 + 6
+3 + (−4)	positive 3 plus negative 4	3 + (−4)
−2 + (−3)	negative 2 plus negative 3	−2 + (−3)

(**Note:** A positive sign indicates either addition *or* a positive number. A negative sign indicates either subtraction *or* a negative number.)

You will use number arrows to help learn the rules for adding signed numbers.

Rule for adding numbers that have the same sign:

Add the numbers and give the sum the same sign as the numbers.

Example 1, though simple, shows the use of number arrows.

EXAMPLE 1   Add +3 and +2.

> **STEP 1**   Add the numbers: 3 + 2 = 5
>
> **STEP 2**   Give the sum a positive sign.
>
> > +3 and +2 = +5
> >
> > or 3 + 2 = 5
>
> **Adding Number Arrows**
>
> Add a +3 arrow and a +2 arrow.

The sum of two or more positive numbers is positive.

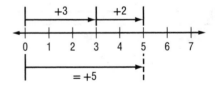

ANSWER: **+5 or 5**

EXAMPLE 2   Add −1 and −4.

> **STEP 1**   Add the numbers: 1 + 4 = 5
>
> **STEP 2**   Give the answer a negative sign.
>
> > −1 + (−4) = −5
>
> **Adding Number Arrows**
>
> Add a −1 arrow and a −4 arrow.

The sum of two or more negative numbers is negative.

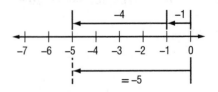

ANSWER: **−5**

Write the addition shown by the number arrows. The first problem is done as an example.

1. <u>4 + 3 = 7</u>

2. _____

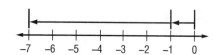

3. _____

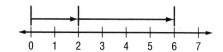

4. _____

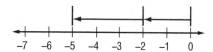

Draw number arrows to represent each addition.

5. $4 + 1$

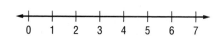

6. $-2 + (-3)$

7. $1 + 4$

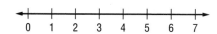

8. $-3 + (-2)$

Add.

9. $5 + 6 =$ $+8 + (+3) =$ $12 + 7 =$ $+9 + (+7) =$

10. $-3 + (-4) =$ $-5 + (-2) =$ $-6 + (-1) =$ $-8 + (-4) =$

11. $-12 + (-9) =$ $-15 + (-6) =$ $-13 + (-12) =$ $-24 + (-17) =$

The sum of a positive number and a negative number may be positive, negative, or zero.

EXAMPLE 3   Add +5 and −2.

> **STEP 1**   Find the difference between 5 and 2.
>
> $5 - 2 = 3$
>
> **STEP 2**   Give the result a positive sign (5 is greater than 2).

Rule for adding a positive and a negative number:

Find the difference between the two numbers and give the result the sign of the greater number.

Add a +5 arrow and a −2 arrow.

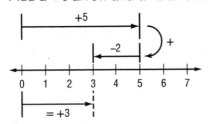

ANSWER: **+3**

(**Note:** Adding a negative number is the same as subtracting a positive number. $5 + (−2) = 5 − (+2) = 3$)

EXAMPLE 4   Add + 2 and −4.

> **STEP 1**   Find the difference between 4 and 2.
>
> $4 - 2 = 2$
>
> **STEP 2**   Give the answer a negative sign (4 is greater than 2).

Add a +2 arrow and a −4 arrow.

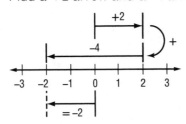

ANSWER: **−2**

EXAMPLE 5   Add −3 and +3.

Find the difference between 3 and 3.

$3 - 3 = 0$

Add a −3 arrow and a +3 arrow.

ANSWER: **0**

Two numbers that differ only by sign are called **opposites.** The sum of opposites is always 0.

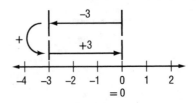

Write the addition shown by the number arrows.

12. _____

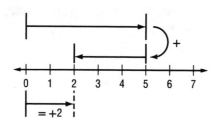

13. _____

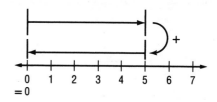

14. _____

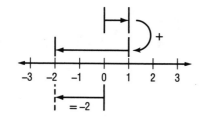

15. _____

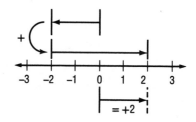

Draw number arrows to represent each addition.

16. $6 + (-4)$

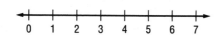

17. $-5 + (+2)$

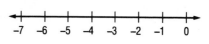

Add.

18. $6 + (-3) =$ $4 + (-7) =$ $+4 + (-4) =$ $+5 + (-6) =$

19. $-3 + 4 =$ $-6 + 3 =$ $-5 + (+9) =$ $-8 + 1 =$

20. $-2 + (+2) =$ $-10 + 5 =$ $13 + (-8) =$ $20 + (-11) =$

Subtracting Signed Numbers

Rule for subtracting signed numbers:
Change the sign of the number being subtracted, and then add the numbers.

EXAMPLE 1 Subtract +5 from +8.

STEP 1 Write $+8 - (+5)$, or $8 - 5$.

STEP 2 Change the sign of +5 to –5, and then add.

$8 - 5 = 8 + (-5) = \mathbf{3}$

In algebra, subtracting a signed number is the same as adding the same number with its sign reversed.

ANSWER: 3

(**Note:** You should continue to subtract positive numbers in the usual way! Example 1 just shows you a new way to think about subtraction.)

EXAMPLE 2 Subtract 12 from 3.

STEP 1 Write $3 - 12$.

STEP 2 Change the sign of 12 to –12 and then add.

$3 - 12 = 3 + (-12) = \mathbf{-9}$

In algebra, you can subtract a greater number from a lesser number! Signed numbers make this possible.

ANSWER: –9

EXAMPLE 3 Subtract –4 from 7.

STEP 1 Write $7 - (-4)$.

STEP 2 Change the sign of –4 to +4 and then add.

$7 - (-4) = 7 + 4 = \mathbf{11}$

Subtracting signed numbers can be thought of as finding the distance between two points on a number line.

ANSWER: 11

Count the units between –4 and +7. Do you get 11?

Circle the subtraction that is represented on each number line.

1.

   a. $-4 - (-5)$
   b. $5 - (-4)$
   c. $5 - 4$
   d. $-5 - 4$

2.

   a. $2 - (-3)$
   b. $-3 - (-2)$
   c. $2 - (+3)$
   d. $-3 - 2$

Subtract.

3. $9 - (+3) =$ $12 - (+7) =$ $8 - 4 =$ $15 - 7 =$

4. $4 - (+8) =$ $7 - 2 =$ $+5 - (+9) =$ $10 - 4 =$

5. $3 - (-2) =$ $8 - (-2) =$ $11 - (-7) =$ $14 - (-9) =$

6. $-7 - (-1) =$ $-9 - (+3) =$ $-4 - (-3) =$ $-1 - (+5) =$

Solve.

7. At 9:00 P.M. the temperature in Chicago was −6°F. By morning the temperature was +7°F. How many degrees did the temperature rise overnight?

8. On January 23, 1971, the temperature in Prospect Creek, Alaska, fell to −79°F. How many degrees below freezing (32°F) did the temperature fall?

9. Mount McKinley in Alaska, the highest mountain in North America, is 20,320 feet above sea level. Death Valley, the lowest point, is 280 feet below sea level. What is the difference in height between Mount McKinley and Death Valley?

10. Mount Whitney in California, the highest mountain in the U.S., excluding Alaska, is 14,495 feet high. What is the difference in height between Mount Whitney and Death Valley? (See problem 9.)

Multiplying Signed Numbers

In algebra, as well as arithmetic, multiplication can be shown in three ways:

- by a times sign: 5×4
- by a dot: $5 \cdot 4$
- by parentheses: $5(4)$

> **Rules for multiplying signed numbers:**
>
> - If the signs of the numbers are alike, multiply the numbers and give the product a positive sign.
>   positive × positive = positive   negative × negative = positive
> - If the signs of the numbers are different, multiply the numbers and give the product a negative sign.
>   positive × negative = negative   negative × positive = negative

 Multiply +4 times +3.   Multiply: $+4 \times +3 = \textbf{+12}$

  STEP 1 Multiply the numbers. $4 \times 3 = 12$

  STEP 2 Because the signs are both positive, the product is positive.

 Find the product of −5 times −6.   Multiply: $-5(-6) = \textbf{+30}$

  STEP 1 Multiply the numbers. $5(6) = 30$

  STEP 2 Because the signs are both negative, the product is positive.

EXAMPLE 3 **Multiply +2 times −7.**   Multiply: $+2(-7) = \textbf{−14}$

  STEP 1 Multiply the numbers. $2(7) = 14$

  STEP 2 Because the signs are different (one positive, one negative), the product is negative.

EXAMPLE 4 **Find the product of −9 times +6.**   Multiply: $-9 \cdot +6 = \textbf{−54}$

  STEP 1 Multiply the numbers. $9 \cdot 6 = 54$

  STEP 2 Because the signs are different (one negative, one positive), the product is negative.

Multiply. (Remember: Like signs give positive products. Unlike signs give negative products.)

1. $3 \cdot 6 =$ $+7 \cdot -5 =$ $+8 \cdot +4 =$ $-6 \cdot -9 =$

2. $-8 \times -7 =$ $5 \times 4 =$ $6 \times -8 =$ $-3 \times 4 =$

3. $(+6)(+6) =$ $(-4)(-7) =$ $7(6) =$ $(-2)(-9) =$

4. $(8)(-5) =$ $(-3)(+3) =$ $(9)(-4) =$ $(-5)(6) =$

5. $(-9)(-5) =$ $(-9)(+6) =$ $(-7)(-3) =$ $(+6)(-4) =$

When you multiply more than two signed numbers, multiply them two at a time or use the shortcut given by the following rule.

Rules for multiplying more than two signed numbers:
Multiply all the numbers together.
- If there is an even number of negative signs, give the product a positive sign.
- If there is an odd number of negative signs, give the product a negative sign.

Multiply. Two problems are done as examples.

6. $(-5)(+2)(-3) = +30$ $(-2)(-2)(+4) =$ $(-2)(+3)(-2)(+4) =$
two negative signs

7. $(-3)(-4)(-2) = -24$ $(-1)(-3)(-6) =$ $(-1)(-3)(-2)(-2)(-3) =$
three negative signs

8. $-3 \cdot -4 =$ $-5 \cdot -2 \cdot -4 =$ $-1 \cdot -4 \cdot -6 \cdot -2 =$

9. $-2 \times -4 \times -2 =$ $-3 \times 5 \times -2 =$ $-1 \times -4 \times 2 =$

Dividing Signed Numbers

Division can also be shown in three ways:
- by a division sign: $15 \div 3$
- by a slash: $15/3$
- by a bar: $\frac{15}{3}$

Rules for dividing signed numbers:

- If the signs of the numbers are alike, divide the numbers and give the quotient a positive sign.

  positive ÷ positive = positive   negative ÷ negative = positive

- If the signs of the numbers are different, divide the numbers and give the quotient a negative sign.

  positive ÷ negative = negative   negative ÷ positive = negative

EXAMPLE 1 Divide +24 by +6.   Divide: $+24 \div +6 = \mathbf{+4}$

STEP 1 Divide the numbers. $24 \div 6 = 4$

STEP 2 Because the signs are both positive, the quotient is positive.

EXAMPLE 2 Find the quotient of −25 divided by −5.   Divide: $\frac{-25}{-5} = \mathbf{+5}$

STEP 1 Divide the numbers. $\frac{25}{5} = 5$

STEP 2 Because the signs are both negative, the quotient is positive.

EXAMPLE 3 Divide +30 by −10.   Divide: $+30 \div -10 = \mathbf{-3}$

STEP 1 Divide the numbers. $30 \div 10 = 3$

STEP 2 Because the signs are different (one positive, one negative), the quotient is negative.

EXAMPLE 4 Find the quotient of −16 divided by +4.   Divide: $\frac{-16}{+4} = \mathbf{-4}$

STEP 1 Divide the numbers. $\frac{16}{4} = 4$

STEP 2 Because the signs are different (one negative, one positive), the quotient is negative.

Divide. (Remember: Like signs give positive quotients. Unlike signs give negative quotients.)

1. $\dfrac{-24}{+4} =$ $\dfrac{+30}{-6} =$ $\dfrac{-42}{-7} =$ $\dfrac{+63}{+9} =$

2. $\dfrac{-104}{-8} =$ $\dfrac{-68}{2} =$ $\dfrac{54}{3} =$ $\dfrac{70}{-5} =$

3. $-18 \div 6 =$ $72 \div -8 =$ $81 \div 9 =$ $-48 \div -6 =$

4. $100 \div -4 =$ $-28 \div 2 =$ $-150 \div -5 =$ $96 \div -4 =$

5. $12/-6 =$ $-4/2 =$ $-18/-3 =$ $-28/7 =$

6. $42/-3 =$ $-60/4 =$ $-72/-6 =$ $-125/5 =$

Sometimes, the answer to a division problem is a fraction. The fraction should be simplified to lowest terms and the sign placed in front of the fraction.

Divide. The first problem in each row is done as an example.

7. $-6 \div 9 = \dfrac{-6}{9} = -\dfrac{2}{3}$ $-9 \div 12 =$ $-5 \div 10 =$ $-8 \div 20 =$

8. $-3 \div -6 = \dfrac{-3}{-6} = \dfrac{1}{2}$ $-4 \div -10 =$ $-12 \div -16 =$ $-8 \div -12 =$

9. $\dfrac{5}{-25} = \dfrac{-5}{25} = -\dfrac{1}{5}$ $\dfrac{-50}{125} =$ $\dfrac{35}{-100} =$ $\dfrac{-100}{150} =$

Applying Your Skills: Signed Numbers

1. During the last decade, the population growth rate of Springfield averaged 2.4% per year. So far this decade, the population growth rate has averaged −0.6%. What has been the change in the average population growth rate?

2. The top of Avery Hill is 328 feet above the entrance to Avery Cave. The lowest point in the cave is 152 feet below the cave entrance. What is the difference in elevation between the top of the mountain and the lowest point in the cave?

3. Measured from sea level, an underwater rock shelf is at a depth of 800 feet. The nearby ocean bottom is at 3 times the depth of the rock shelf.
   a. What is the depth of the nearby ocean bottom?
   b. Express your answer to **part a** as an elevation where positive numbers mean *above sea level* and negative numbers mean *below sea level*.

4. The stock price of AM Company was 47.35 on Monday morning. At the end of the day, the price had changed by −0.62. What was the price of the stock at the end of the day?

5. You are hired as a bookkeeper for a small company. You discover that the previous bookkeeper wrote negative numbers to represent bills (amount owned). You see the following entries:

Emerald Gravel	−$342.50
Valley Tile	−$128.00
William's Paint	−$84.50
Crowe Printing Co.	−$263.50

   What is total amount owed?

6. The temperatures in Chicago at 6:00 P.M. and at midnight are shown on the two thermometers at the right. What was the change in temperature between 6:00 P.M. and midnight?

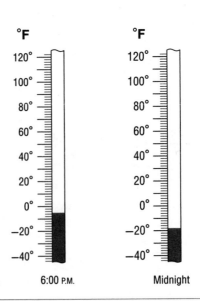

7. Cascade Mining is drilling a mine ventilation shaft. The mine is at a depth of 630 ft. The drill has reached $\frac{1}{3}$ of the way to the mine.

   a. At what depth is the drill at this point?

   b. Express your answer to **part a** as a negative number, where a negative number means *feet below the surface.*

8. Saturday, following a storm, the Mississippi River was 3 feet above flood level. By Wednesday, the water level had changed by −4 feet. Compared to flood level, what was the height of the water on Wednesday? (Write your answer as a positive or a negative number.)

9. Today's wind makes the temperature feel twice as cold as the temperature indicated on the thermometer. If the thermometer reads −12°, how cold does the wind make it feel?

10. Alan Construction is lowering the surface of a lot in preparation for a new building. So far, 200 truckloads of dirt have been removed to bring the level of the surface to −4 feet. How many *more* truckloads of dirt will need to be removed to lower the ground level to −10 feet?

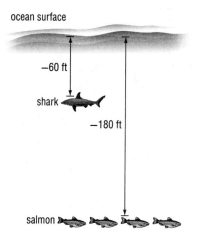

11. As shown at the right, a school of salmon is swimming below a shark.

    a. What is the distance between the shark and the salmon?

    b. What is the ratio of the depth at which the salmon are swimming to the depth at which the shark is swimming?

12. Angie has been dieting for several months. Her monthly pounds-gained (+) or pounds-lost (−) results are shown in the table below.

Month	Result
Feb.	−4
Mar.	−6
Apr.	+1
May	+3
June	−4
July	−8

   a. How many pounds did Angie lose during the six months shown?

   b. What was the average monthly change in Angie's weight?

Signed Numbers Review

Solve the problems below. When you finish, check your answers at the back of the book. Then correct any errors.

Write each number as a signed number with a + sign or a − sign.

1. positive 13   positive $2\frac{1}{2}$ negative 26   negative $\frac{7}{8}$

Locate these points on the number line. $-10, -7, -4\frac{1}{2}, -2, -\frac{1}{4}, 1\frac{1}{2}, 5, 8, 9\frac{3}{4}$

2.

Write the addition shown by each pair of number arrows.

3. _____

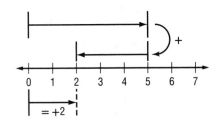

4. _____

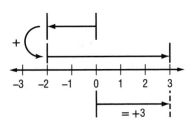

Add.

5. $7 + (-2) =$ $-6 + 9 =$ $-5 + (+5) =$ $11 + (-2) =$

6. $-8 + (-4) =$ $-9 + (-3) =$ $-6 + (+9) =$ $-7 + 3 =$

Subtract.

7. $8 - (+3) =$ $6 - (+6) =$ $5 - (+8) =$ $12 - (-7) =$

8. $-5 - (-7) =$ $-11 - 2 =$ $-5 - (+3) =$ $-9 - (-3) =$

Solve.

9. At 10:00 A.M. the temperature in New York City was $-6°$F. At 3:00 P.M. the temperature had risen to $+17°$F. How many degrees did the temperature rise?

Multiply.

10. $(9)(-3) =$ $(-4)(+7) =$ $(6)(-3) =$ $(-8)(2) =$

11. $(-6)(-5) =$ $(+5)(+3) =$ $(-8)(-4) =$ $(9)(6) =$

12. $(-2)(-4)(-3) =$ $(-2)(+3)(-6) =$ $(-1)(-3)(-2)(-3) =$

Divide.

13. $\dfrac{-30}{+6} =$ $\dfrac{+28}{-4} =$ $\dfrac{-48}{-6} =$ $\dfrac{+56}{+7} =$

14. $24 \div -6 =$ $-64 \div -8 =$ $49 \div -7 =$ $-54 \div 9 =$

15. $18/-3 =$ $-14/7 =$ $-24/-8 =$ $-27/3 =$

16. $-14 \div -16 =$ $8 \div -10 =$ $\dfrac{-6}{8} =$ $\dfrac{-10}{-15} =$

POWERS AND ROOTS

What Is a Power?

A **power** is the product of a number multiplied by itself one or more times. *Four to the third power* means $4 \times 4 \times 4$.

A power is commonly written as a **base** and an **exponent.**

$4 \times 4 \times 4$ is written 4^3 ← exponent

$\uparrow$

base

The 4 is called the *base*. The base is the number being multiplied.

The 3 is called the *exponent*. The exponent tells how many times the base is written in the product.

Look at these examples.

Product	As a base and exponent	In words
$3 \cdot 3$	3^2	3 to the second power *or* 3 squared
$(-4)(-4)(-4)$	$(-4)^3$	negative 4 to the third power *or* negative 4 cubed
$\frac{1}{2} \times \frac{1}{2} \times \frac{1}{2} \times \frac{1}{2}$	$\left(\frac{1}{2}\right)^4$	$\frac{1}{2}$ to the fourth power

(**Note:** A number raised to the second power is often called *squared* and raised to the third power is often called *cubed*. These are the only two powers that have special names.)

The **value** of a power is found by multiplication.

EXAMPLE 1 **Find the value of 4^4.**

STEP 1 Write the power 4^4 as a product. $\qquad$ $4^4 = (4)(4)(4)(4)$

STEP 2 Multiply the first two terms together. $\qquad$ $= (16)(4)(4)$
$(4)(4) = 16$

STEP 3 Multiply the answer in Step 2 times the $\qquad$ $= (64)(4)$
next number. $(16)(4) = 64$

STEP 4 Continue multiplying until you have used $\qquad$ $= \mathbf{256}$
all the terms. $(64)(4) = 256$

ANSWER: **256**

EXAMPLE 2 Find the value of $\left(-\frac{2}{3}\right)^3$.

     STEP 1 Write the power $\left(-\frac{2}{3}\right)^3$ as a product. $\left(-\frac{2}{3}\right)^3 = \left(-\frac{2}{3}\right)\left(-\frac{2}{3}\right)\left(-\frac{2}{3}\right)$

     STEP 2 Multiply the first two terms. $\left(-\frac{2}{3}\right)\left(-\frac{2}{3}\right) = +\frac{4}{9}$ $= \left(+\frac{4}{9}\right)\left(-\frac{2}{3}\right)$

     STEP 3 Multiply the remaining terms. $= -\frac{8}{27}$

ANSWER: $-\frac{8}{27}$

Be sure that the sign of your answer is correct.

Complete the following chart.

Product	Base/Exponent	In Words
1. $(-1)(-1)$	$(-1)^2$	-1 to the second power or -1 squared
2. $(+5)(+5)$		
3. $\left(+\frac{1}{2}\right)\left(+\frac{1}{2}\right)$		
4. 6×6		
5. $7 \cdot 7 \cdot 7$		7 to the third power or 7 cubed
6. $(-3)(-3)(-3)$		
7. $(6)(6)(6)$		
8. $\left(-\frac{3}{4}\right)\left(-\frac{3}{4}\right)\left(-\frac{3}{4}\right)$		
9. $2 \times 2 \times 2 \times 2$		2 to the fourth power
10. $5 \cdot 5 \cdot 5 \cdot 5$		
11. $(-4)(-4)(-4)(-4)$		
12. $\left(\frac{1}{4}\right)\left(\frac{1}{4}\right)\left(\frac{1}{4}\right)\left(\frac{1}{4}\right)$		

Find the value of each power.

13. $5^2 =$ $(-2)^2 =$ $(9)^2 =$ $(-6)^2 =$

14. $5^3 =$ $(3)^4 =$ $(-10)^3 =$ $(-3)^4 =$

15. $\left(\frac{2}{3}\right)^2 =$ $\left(-\frac{4}{5}\right)^3 =$ $\left(\frac{1}{2}\right)^4 =$ $\left(\frac{3}{4}\right)^3 =$

There are two special cases of powers.

> **Rule:** Any number raised to the first power is that number.

EXAMPLE 3 $8^1 = 8$

An exponent of 1 means the base 8 is written only once and is not multiplied.

> **Rule:** Any number raised to the 0 power is 1.

EXAMPLE 4 $6^0 = 1$

This is the hardest rule to remember. A zero exponent means *a number divided by itself*.

To simplify an expression containing more than one power, find the value for each power and then follow the rules for adding and subtracting signed numbers.

EXAMPLE 5 **Simplify the expression $(-4)^3 - 3^4 + 7^0$.**

STEP 1 Find the value of each power.

$(-4)^3 = -64$; $3^4 = 81$; $7^0 = 1$

STEP 2 Substitute the values and follow the rules for adding and subtracting signed numbers.

$(-4)^3 - 3^4 + 7^0 = -64 - 81 + 1$

$= -64 + (-81) + 1$ (Adding −81 is the same as subtracting 81.)

$= -145 + 1$

$= \mathbf{-144}$

ANSWER: **−144**

Find the value of each expression.

16. $8^2 - 3^1 =$ $9^2 - 2^0 =$ $7^0 + 4^1 =$

17. $5^0 + (-3)^2 + 4^1 =$ $4^3 - 3^2 + 5^0 =$ $(-2)^3 - (-3)^3 + 2^1 =$

18. $\left(\frac{1}{2}\right)^2 + \left(\frac{1}{2}\right)^1 =$ $\left(-\frac{3}{4}\right)^2 - \left(\frac{1}{2}\right)^1 =$ $\left(\frac{3}{4}\right)^2 + \left(-\frac{1}{2}\right)^1 + \left(\frac{1}{2}\right)^0 =$

Multiplication and Division of Powers

You may need to find the value of a product or quotient of powers. As your first step, find the value of each power and then multiply or divide as indicated.

EXAMPLE 1   Find the value of the product $3^2 \cdot 2^4$.

STEP 1   Find the value of each power.

$3^2 = 9$ and $2^4 = 16$

STEP 2   Multiply the values found in Step 1.

$3^2 \cdot 2^4 = 9 \cdot 16 = \mathbf{144}$

ANSWER: **144**

EXAMPLE 2   Find the value of the quotient $\frac{4^3}{2^2}$.

STEP 1   Find the value of each power.

$4^3 = 64$ and $2^2 = 4$

STEP 2   Divide the values found in Step 1.

$\frac{4^3}{2^2} = \frac{64}{4} = \mathbf{16}$

ANSWER: **16**

Find the value of each product or quotient and simplify your answer. The first problem in each row is done as an example.

1. $2^3 \cdot 4^2 = 8 \cdot 16$
   $= 128$

     $4^2 \cdot 3^3 =$

     $(-2)^3 \cdot 5^2 =$

2. $\left(\frac{1}{2}\right)^3 \times 5^2 = \left(\frac{1}{8}\right) \times 25$
   $= 3\frac{1}{8}$

     $\left(\frac{1}{5}\right)^2 \times 5^3 =$

     $\left(\frac{1}{4}\right)^2 \times \left(\frac{4}{3}\right)^3 =$

3. $\frac{5^3}{7^2} = \frac{125}{49}$
   $= 2\frac{27}{49}$

     $\frac{3^4}{7^2} =$

     $\frac{(-4)^3}{2^4} =$

To solve more complicated problems, first find the value of each power. Then simplify when possible.

4. $\frac{8^2 \times 4^1}{2^3 \times 3^2} = \frac{\overset{8}{\cancel{64}} \times 4}{\cancel{8}_1 \times 9}$
   $= \frac{8 \times 4}{9}$
   $= \frac{32}{9}$
   $= 3\frac{5}{9}$

     $\frac{7^2 \times 2^3}{4^2 \times 3^2} =$

     $\frac{4^2 \times 5^1}{2^3 \times (-6)^2} =$

Simplifying Products and Quotients with Like Bases

Products of Powers with Like Bases

Multiplication often involves **like bases** (same base numbers). As the examples show, a product of powers with like bases can be simplified by adding exponents.

EXAMPLE 1 Simplify $3^4 \times 3^2$.

One way to simplify this product is to write each power as a product and then count 3s.

$3^4 \times 3^2 = 3 \times 3 \times 3 \times 3 \times 3 \times 3 = \mathbf{3^6}$ (There are six 3s in the product.)

ANSWER: $\mathbf{3^6}$

A shortcut is to add the exponents of each power.

$3^4 \times 3^2 = 3^{4+2} = \mathbf{3^6}$

Seeing this, you can write the rule for simplifying a product of powers with like bases.

Rule for simplifying a product of powers with like bases:
To multiply numbers with the same base, add the exponents.

EXAMPLE 2 Simplify $2^2 \cdot 2^3$.

$2^2 \cdot 2^3 = 2^{2+3} = \mathbf{2^5}$

ANSWER: $\mathbf{2^5}$

EXAMPLE 3 Simplify $(4^3)(4^2)(4^1)$.

$(4^3)(4^2)(4^1) = 4^{3+2+1} = \mathbf{4^6}$

ANSWER: $\mathbf{4^6}$

Add exponents to simplify each product of powers. The first problem in each row is done as an example.

1. $2^3 \times 2^4 = \mathbf{2^{3+4}} = \mathbf{2^7}$ $3^2 \times 3^2 =$ $4^4 \times 4^3 =$

2. $5^2 \cdot 5^1 = \mathbf{5^{2+1}} = \mathbf{5^3}$ $7^3 \cdot 7^1 =$ $4^3 \cdot 4^1 =$

3. $(6^2)(6^1)(6^0) = \mathbf{6^{2+1+0}}$ $(2^2)(2^3)(2^0) =$ $(8^3)(8^2)(8^2) =$
   $= \mathbf{6^3}$

4. $4^3 \cdot 4^2 \cdot 3^4 = \mathbf{4^{3+2}} \cdot \mathbf{3^4}$ $(5^7)(5^5)(2^3) =$ $7^4 \times 5^3 \times 7^2 =$
   $= \mathbf{4^5} \cdot \mathbf{3^4}$
   (4 and 3 are not like bases)

Quotients of Powers with Like Bases

Division also may involve like bases. A quotient of powers with like bases can be simplified by subtracting exponents.

EXAMPLE 4 Simplify the quotient $4^7 \div 4^5$.

The long way is to write each power as a product and then simplify by dividing out the 4s.

$$\frac{4^7}{4^5} = \frac{\cancel{4} \times \cancel{4} \times \cancel{4} \times \cancel{4} \times \cancel{4} \times 4 \times 4}{\cancel{4} \times \cancel{4} \times \cancel{4} \times \cancel{4} \times \cancel{4}} = 4 \times 4 = \mathbf{4^2} \text{ (There are two 4s left in the product.)}$$

ANSWER: $\mathbf{4^2}$

A shortcut is to subtract the exponents of each power.

$$\frac{4^7}{4^5} = 4^{7-5} = \mathbf{4^2}$$

Seeing this, you can write the rule for simplifying a quotient of powers with like bases.

> **Rule for simplifying a quotient of powers with like bases:**
> To divide numbers with the same base, subtract the exponents.

EXAMPLE 5 Simplify $2^6 \div 2^3$.

$$2^6 \div 2^3 = 2^{6-3} = \mathbf{2^3}$$

ANSWER: $\mathbf{2^3}$

EXAMPLE 6 Simplify $7^4 \div 7^3$.

$$7^4 \div 7^3 = 7^{4-3} = 7^1 = \mathbf{7}$$

ANSWER: 7

Subtract exponents to simplify each quotient of powers. The first problem in each row is done as an example.

5. $5^3 \div 5^1 = \mathbf{5^{3-1}} = \mathbf{5^2}$ $2^6 \div 2^2 =$ $8^3 \div 8^1 =$

6. $7^3 \div 7^2 = \mathbf{7^{3-2}} = \mathbf{7^1} = \mathbf{7}$ $8^4 \div 8^3 =$ $5^7 \div 5^6 =$

7. $\dfrac{4^7}{4^5} = \mathbf{4^{7-5}} = \mathbf{4^2}$ $\dfrac{8^6}{8^3} =$ $\dfrac{7^6}{7^5} =$

8. $\dfrac{6^5 \times 4^2}{6^2} = \mathbf{6^{5-2} \times 4^2}$
    $\phantom{\dfrac{6^5 \times 4^2}{6^2}} = \mathbf{6^3 \times 4^2}$

    (6 and 4 are not like bases) $\dfrac{10^9 \times 5^4}{10^7} =$ $\dfrac{9^5 \times 7^6}{9^3 \times 7^3} =$

Negative Exponents

Negative numbers are also used as exponents. A **negative exponent** stands for the number found by inverting a power. For example, $4^{-2} = \frac{1}{4^2} = \frac{1}{4 \times 4}$. Below are more examples.

Power with Negative Exponent	Meaning	In Words	Value
3^{-1}	$\frac{1}{3}$	3 to the minus 1st power	$\frac{1}{3}$
5^{-2}	$\frac{1}{5 \times 5}$	5 to the minus 2nd power	$\frac{1}{25}$
4^{-3}	$\frac{1}{4 \times 4 \times 4}$	4 to the minus 3rd power	$\frac{1}{64}$
2^{-4}	$\frac{1}{2 \times 2 \times 2 \times 2}$	2 to the minus 4th power	$\frac{1}{16}$

The rules that apply to positive exponents also apply to negative exponents.

Add exponents when multiplying like bases.

EXAMPLE 1 Simplify $3^{-4} \times 3^{-2}$.

$$3^{-4} \times 3^{-2} = 3^{-4 + (-2)} = \mathbf{3^{-6}}$$

ANSWER: $\mathbf{3^{-6}}$

Subtract exponents when dividing like bases.

EXAMPLE 2 Simplify $5^4 \div 5^{-3}$.

$$5^4 \div 5^{-3} = 5^{4-(-3)} = \mathbf{5^7}$$

ANSWER: $\mathbf{5^7}$

	Meaning	In Words	Value
1. 7^{-2}	_____	_____	_____
2. 10^{-3}	_____	_____	_____
3. 3^{-4}	_____	_____	_____

Simplify each product or quotient.

4. $3^3 \times 3^{-1} =$ $4^{-2} \times 4^2 =$ $10^3 \times 10^{-4} =$

5. $6^2 \div 6^{-1} =$ $2^4 \div 2^{-2} =$ $5^3 \div 5^{-1} =$

Scientific Notation

Scientific notation is a shorthand way of writing numbers. In scientific notation, a number is written as the product of two factors. The first factor is a number between 1 and 10. The second factor is a power of 10. The power of 10 factor tells how many places to move the decimal point when changing from scientific notation to regular notation.

- A positive exponent means to move the decimal point to the right.
- A negative exponent means to move the decimal point to the left.

Positive Exponents

EXAMPLE 1 $7.5 \times 10^5 = 750,000$

Move the decimal point five places to the right.

Scientific Notation	Number
3.8×10^3	3,800
8.5×10^4	85,000
6×10^7	60,000,000

Negative Exponents

EXAMPLE 2 $4.2 \times 10^{-3} = 0.0042$

Move the decimal point three places to the left.

Scientific Notation	Number
5×10^{-2}	0.05
6×10^{-3}	0.006
7.5×10^{-5}	0.000075

Write each number in regular notation.

1. 6.3×10^2 8.75×10^6 9×10^7

2. 5.1×10^{-2} 8.5×10^{-5} 7×10^{-6}

Using a positive exponent, write each number in scientific notation.

3. 5,000   30,000   75,000,000

Using a negative exponent, write each number in scientific notation.

4. 0.03   0.0075   0.0000125

Solve.

5. The approximate distance from Earth to the Moon is 2.4×10^5 miles. Write this distance as a whole number.

6. The diameter of a human hair is measured to be 0.0012 inches. Write this diameter in scientific notation.

What Is a Square Root?

An important skill is finding the **square root** of a positive number. For example, to find the square root of 25, ask, "What number multiplied by itself equals 25?"

The number 25 has two square roots.

- 5 is a square root of 25. $5^2 = 25$

- −5 is a square root of 25. $(−5)^2 = 25$

The square root symbol is $\sqrt{}$.

Thus, $\sqrt{25} = +5$ and −5, or $\sqrt{25} = \pm 5$.

> A positive number has two square roots.
> - a positive square root (+)
> - a negative square root (−)
> - ± means "both positive and negative."
>
> EXAMPLE $\sqrt{36} = \pm 6$

EXAMPLE 1 **7 is the square root of what number?**

Since $7^2 = 49$, 7 is the square root of 49.

ANSWER: $7 = \sqrt{49}$

EXAMPLE 2 **Find $\sqrt{64}$.**

$8^2 = 64$ and $(−8)^2 = 64$.

ANSWER: $\sqrt{64} = \pm 8$

Numbers that have whole number square roots are called **perfect squares.** Perfect squares are easily found by squaring whole numbers. The first fifteen perfect squares are shown in the table below.

Table of Perfect Squares		
$1^2 = 1$	$6^2 = 36$	$11^2 = 121$
$2^2 = 4$	$7^2 = 49$	$12^2 = 144$
$3^2 = 9$	$8^2 = 64$	$13^2 = 169$
$4^2 = 16$	$9^2 = 81$	$14^2 = 196$
$5^2 = 25$	$10^2 = 100$	$15^2 = 225$

As you already know, 0 times 0 equals 0. Therefore, $0^2 = 0$. To find the square root of a perfect square from the table,
- find the perfect square in the right-hand column,
- read the whole number square root in the left-hand column.

(**Remember:** Each perfect square has both a positive and a negative square root. For example, $\sqrt{121} = \pm 11$)

From the table on the previous page, find the square root of each perfect square below.

1. $\sqrt{169} = \pm 13$ $\sqrt{25} =$ $\sqrt{81} =$

2. $\sqrt{4} =$ $\sqrt{121} =$ $\sqrt{36} =$

3. $\sqrt{49} =$ $\sqrt{1} =$ $\sqrt{225} =$

4. $\sqrt{144} =$ $\sqrt{64} =$ $\sqrt{9} =$

5. $\sqrt{100} =$ $\sqrt{16} =$ $\sqrt{196} =$

The square root of a fraction is equal to the square root of the numerator over the square root of the denominator.

EXAMPLE 3 Find $\sqrt{\frac{36}{49}}$.

STEP 1 Write as the square root of the numerator over the square root of the denominator. $\sqrt{\frac{36}{49}} = \frac{\sqrt{36}}{\sqrt{49}}$

STEP 2 Find the square roots of both the numerator and the denominator. $= \pm\frac{6}{7}$

ANSWER: $\pm\frac{6}{7}$ **Check:** $\left(\frac{6}{7}\right)^2 = \frac{36}{49}$ and $\left(-\frac{6}{7}\right)^2 = \frac{36}{49}$

Find the square root of each fraction below. The first problem is done as an example.

6. $\sqrt{\frac{4}{9}} = \pm\frac{2}{3}$ $\sqrt{\frac{9}{64}} =$ $\sqrt{\frac{25}{36}} =$

7. $\sqrt{\frac{1}{4}} =$ $\sqrt{\frac{49}{81}} =$ $\sqrt{\frac{121}{144}} =$

8. $\sqrt{\frac{81}{16}} =$ $\sqrt{\frac{1}{196}} =$ $\sqrt{\frac{100}{121}} =$

Finding an Approximate Square Root

A number that is not a perfect square does not have a whole number square root. For example, the square root of 30 is between the whole numbers 5 and 6.

$$\sqrt{30} \text{ is greater than 5, since } 5^2 = 25$$

$$\sqrt{30} \text{ is less than 6, since } 6^2 = 36$$

To find an approximate square root of a number that is not a perfect square, follow these three steps.

STEP 1 Choose a perfect square that is close to the square root you want to find.

STEP 2 Divide this chosen number into the number you are trying to find the square root of.

STEP 3 Average your choice from Step 1 with the answer from Step 2. The average of these two numbers is an approximate square root.

(**Note:** The $\approx$ sign means "is approximately equal to.")

EXAMPLE 1 **Find an approximate positive square root of 30.**

STEP 1 Find the perfect square that is closest to the number 30. This perfect square is 25.

$5 \approx \sqrt{30}$
because $5 = \sqrt{25}$

STEP 2 Divide 5 into 30.

$30 \div 5 = 6$

STEP 3 Find the average of the numbers 5 and 6. The average is the approximate square root of 30.

$5 + 6 = 11$
$11 \div 2 = 5\frac{1}{2}$

ANSWER: $5\frac{1}{2}$ or 5.5

Check: $(5.5)^2 = 30.25$

EXAMPLE 2 **Find an approximate positive square root of 54.**

STEP 1 Find the perfect square that is closest to the number 54. Choose $7 \approx \sqrt{54}$ because $7 = \sqrt{49}$.

$7 \approx \sqrt{54}$

STEP 2 Divide 7 into 54.

$54 \div 7 = 7.7$

STEP 3 Find the average of the numbers 7 and 7.7.

$7 + 7.7 = 14.7$
$14.7 \div 2 \approx 7.4$

ANSWER: 7.4

Check: $(7.4)^2 = 54.76$

Find an approximate positive square root for each number. Round decimals to the nearest tenth at each step.

1. $\sqrt{42} \approx$

2. $\sqrt{19} \approx$

3. $\sqrt{76} \approx$

4. $\sqrt{90} \approx$

5. $\sqrt{28} \approx$

6. $\sqrt{50} \approx$

7. $\sqrt{137} \approx$

8. $\sqrt{208} \approx$

Applying Your Skills: Powers and Roots

1. A checkerboard has 8^2 squares. Written as a whole number, how many squares are on a checkerboard?

2. Suppose a rock is dropped from a bridge. The distance d the rock will fall in a time t is given by the formula $d = 16t^2$, where d is in feet and t is in seconds.

   a. In the first second, how far will the rock fall?

   b. After two seconds, how far will the rock have fallen?

3. You work for a company that makes toy marbles. The radius of the most popular marble is 0.3 inches. The volume of a marble is given by the formula $V = \frac{4}{3}\pi r^3$, where $\pi \approx 3.14$.

   a. Rounded to the nearest tenth of a cubic inch, what is the volume of glass needed to make one of the most popular marbles?

   b. What volume of glass is needed to make 10^5 marbles?

4. You work in a school learning center and want to help students understand exponents. A student asks for help finding the value of the following expression: $3^2 - 2^3$.

   What would you tell this student?

5. Using a sensitive scale, a lab technician finds that a single grain of pollen weighs about 4.2×10^{-2} grams. What is the approximate weight of 10^6 grains?

6. As a teacher's aide, you want to help students realize just how small an atom is. You find out two facts.

   1) A piece of thin window glass is about 10^{-1} inches thick.

   2) An atom has a diameter of about 10^{-8} inches.

   About how many atoms, placed side by side, would it take to span the thickness of a piece of window glass?

7. A 1-inch thick stack of printer paper is found to contain 175 sheets.

   a. To the nearest thousandth of an inch, how thick is each sheet?

   b. Write this thickness in scientific notation.

   c. Written in scientific notation, what would be the approximate thickness of 20 sheets?

8. A forester's assistant looks at a high-altitude photograph to estimate the number of trees per acre of land in a national forest. She estimates there are 4^3 trees per square inch of the photograph. If an acre is represented by 4^2 square inches, about how many trees are in each acre?

9. You are a quality-control operator at an electronics company. You are told to expect only one defective switch out of every 3^5 switches produced. How many defective switches can you expect out of every 3^8 produced?

10. Sunlight travels from the Sun to Earth at a speed of about 1.9×10^5 miles per second. The distance from the Sun to Earth is about 9.3×10^7 miles.

    a. Divide distance by speed to find the number of seconds it takes sunlight to travel from the Sun to Earth. (Divide: $\frac{9.3}{1.9} \times \frac{10^7}{10^5}$)

    b. Change your answer in **part a** to the nearest whole minute.

11. A nutritionist tells you that an adult needs about 4.5×10^3 mg of potassium each day. She also tells you that a typical banana contains about 4.5×10^2 mg of potassium. How many bananas would you need to eat daily to fulfill your potassium needs?

12. The length in feet of the diagonal support beam shown at right is given by the formula $l = \sqrt{(7^2 + 10^2)}$.

    What is the approximate length of this diagonal?
    (*Hint*: Add the squares of the two sides. Then, find an approximate square root of the sum.)

Powers and Roots Review

Solve the problems below. When you finish, check your answers at the back of the book. Then correct any errors.

1. In the expression 3^4, _____ is the base and _____ is the exponent.

Write each product as a base and exponent.

2. $(-4)(-4)(-4)(-4) =$

3. $(+5)(+5)(+5) =$

4. $\left(-\frac{3}{5}\right)\left(-\frac{3}{5}\right) =$

5. $8 \cdot 8 =$

6. $\left(\frac{2}{3}\right)\left(\frac{2}{3}\right)\left(\frac{2}{3}\right) =$

7. $9 \times 9 \times 9 \times 9 =$

Find each value.

8. $9^2 =$

9. $(-5)^3 =$

10. $(-3)^4 =$

11. $3^0 =$

12. $7^1 =$

13. $\left(\frac{2}{3}\right)^2 =$

14. $5^2 - (-3)^2 + 3^0 =$

15. $\left(\frac{3}{2}\right)^2 + \left(\frac{1}{4}\right)^1 - \left(\frac{1}{2}\right)^0 =$

16. $\frac{4^3}{3^2} =$

17. $\frac{3^2 \times 2^3}{4^2 \times 5^1} =$

Simplify each product or quotient.

18. $(5^4)(5^2)(5^3) =$

19. $\frac{8^6}{8^3} =$

Find each value.

20. $4^{-3} =$

21. $2^4 \cdot 2^{-3} =$

22. $10^{-1} \div 10^2 =$

Write each number in regular notation.

23. 4.5×10^4

24. 2.3×10^{-3}

Write each number in scientific notation.

25. 720,000

26. 0.006

Find each square root.

27. $\sqrt{36} =$

28. $\sqrt{\dfrac{49}{64}} =$

29. Find an approximate square root of 56.

30. Find an approximate square root of 83.

Writing Algebraic Expressions

In algebra, letters called **variables** or **unknowns** are used to represent numbers. A variable may be used in addition, subtraction, multiplication, or division problems. A variable may appear alone or it may be multiplied by a number called a **coefficient.**

$3x + 4 \longleftarrow$ number standing alone

variable

coefficient

An **algebraic expression** consists of variables and numbers combined with one or more of the operations of addition, subtraction, multiplication, or division.

Operation	Algebraic Expression	Word Expression
Addition	$x + 4$	x plus 4
Subtraction	$y - 6$	y minus 6
	or	
	$6 - y$	6 minus y
Multiplication	$5z$	5 times z
Division	$\frac{n}{2}$	n divided by 2
	or	
	$\frac{2}{n}$	2 over n

Write an algebraic expression for each word expression.

1. n minus 8 _____

2. s divided by 5 _____

3. y plus 12 _____

4. w subtract 4 _____

5. 7 times x _____

6. r added to 19 _____

Write a word expression for each algebraic expression.

7. $5x$ _____

8. $4 + n$ _____

9. $7 - n$ _____

10. $\frac{m}{2}$ _____

11. $\frac{a}{5}$ _____

12. $s - 3$ _____

Using Parentheses with Algebraic Expressions

Many algebraic expressions contain more than one operation. The expression $2x + 8$ involves multiplication *and* addition. An algebraic expression may also contain parentheses. Placing a number outside of parentheses means that the whole expression is to be multiplied by the number.

Look at the difference between the expressions $3y + 7$ and $3(y + 7)$ when y is equal to 2.

Without parentheses: $3y + 7 = 3(2) + 7 = 6 + 7 = \mathbf{13}$
With parentheses: $3(y + 7) = 3(2 + 7) = 3(9) = \mathbf{27}$

Parentheses change the meaning and value of an expression. Below are more examples.

Algebraic Expression	Word Expression
$2y - 7$	$2y$ minus 7
$2(y - 7)$	2 times the quantity y minus 7
$\frac{2}{3}x + 5$	$\frac{2}{3}$ x plus 5
$\frac{2}{3}(x + 5)$	$\frac{2}{3}$ times the quantity x plus 5
$3x^2 + 1$	3 times x-squared plus 1
$3(x^2 + 1)$	3 times the quantity x-squared plus 1

Match each algebraic expression with its equivalent word expression.

Algebraic Expressions	Equivalent Word Expressions
_____ **13.** $4x - 9$	**a.** z squared minus 3
_____ **14.** $-2a(a + 7)$	**b.** 4 times the quantity x minus 9
_____ **15.** $z^2 - 3$	**c.** -2 times a squared plus 7
_____ **16.** $\frac{3}{4}x + 4$	**d.** $\frac{3}{4}$ times the quantity x plus 4
_____ **17.** $z(z - 3)$	**e.** $4x$ minus 9
_____ **18.** $4(x - 9)$	**f.** $\frac{3}{4}x$ plus 4
_____ **19.** $-2a^2 + 7$	**g.** z times the quantity z minus 3
_____ **20.** $\frac{3}{4}(x + 4)$	**h.** $-2a$ times the quantity a plus 7

Applying Your Skills: Algebraic Expressions

Algebraic expressions are often part of word problems.

EXAMPLE 1 Suppose *d* represents the amount that Joclyn earns each hour babysitting. Last Saturday Joclyn babysat for 6.5 hours. Write an algebraic expression that represents what Joclyn earned last Saturday.

ANSWER: **6.5*d*** (hours times *d*)

EXAMPLE 2 Mannie and two friends share the cost of a large pizza. Let *c* represent the price of the pizza. Write an algebraic expression that represents Mannie's share of the bill.

ANSWER: $\frac{c}{3}$ (cost divided by 3)

For problems 1–8, choose the correct algebraic expression.

1. On his last birthday, Arnie was 144 centimeters tall. Since that time, he has grown *h* centimeters. Which expression tells Arnie's present height?

    a. $h - 144$ **b.** $144 - h$ **c.** $h + 144$ **d.** $144h$

2. Shelley, Lauren, and Amy are going to share equally the cost of Sunday's breakfast. If the total bill is *t*, which expression tells each person's share?

    a. $\frac{t}{3}$ **b.** $t + 3$ **c.** $3t$ **d.** $t - 3$

3. Mai bought a magazine for \$2.95. If she paid the clerk with *x* dollars, which expression tells how much change Mai would receive?

    a. $\$2.95 - x$ **b.** $\$2.95x$ **c.** $x + \$2.95$ **d.** $x - \$2.95$

4. Blair had a bag of *n* jelly beans. He gave 5 jelly beans to Andre and then divided the remaining candy among three other friends. Which expression tells how many jelly beans each of the other friends got?

    a. $\frac{(n + 5)}{3}$ **b.** $\frac{(n - 5)}{3}$ **c.** $3(n + 5)$ **d.** $3(n - 5)$

5. Christy lost 5% of her body weight while on a diet. If Christy weighed *n* pounds before dieting, which expression tells her weight now?

    a. $n + 0.05$ **b.** $n - 0.05$ **c.** $n + 0.05n$ **d.** $n - 0.05n$

6. If you subtract 6 from the square of Gary's age, the result is Zane's age. If Gary is g years old, which expression tells Zane's age?

   a. $g - 6^2$ **b.** $(g - 6)^2$ **c.** $g^2 - 6$ **d.** $(6 - g)^2$

Problems 7 and 8 are just for fun.

7. If you multiply the square of a number n by 6 and then add 3 to the product, the sum is equal to 57. Which expression below equals 57?

   a. $6n^2 + 3$ **b.** $6(n^2 + 3)$ **c.** $6(n + 3)^2$ **d.** $(6n)^2 + 3$

8. Heidi has several coins. If you subtract 3 from the number she has and then square the difference, the result tells the number of dollar bills she has. If Heidi has n coins, which expression tells how many dollar bills she has?

   a. $n^2 - 3$ **b.** $n(n - 3)^2$ **c.** $(n - 3)^2$ **d.** $(n - 3)^2 - n$

For problems 9–14, write an algebraic expression as indicated.

9. Kelli is 154 centimeters tall. Suppose she grows h centimeters this next year. Write an expression that tells her height at the end of the year.

10. Marcel and four friends agree to share the cost of renting several video games. The rental cost is n dollars. Write an expression that tells Marcel's share.

11. A sweater regularly selling for p dollars is on sale for 30% off. Write an expression that tells the savings being offered on the sweater.

12. Caren bought a jacket on sale. She paid 25% less than original price of m dollars. Write an expression that tells the amount that Caren paid.

13. If you multiply the square of a number n by 3 and then subtract 4 from the product, the difference equals Ben's age. Write an expression that tells Ben's age.

14. Wayne describes an algebraic expression. He says, "Subtract 12 times a number b from twice the square of the number." Write Wayne's expression.

Evaluating Algebraic Expressions

You find the value of an algebraic expression by substituting a number for each variable and then doing the operations in the following order.

1. Evaluate expressions within parentheses.
2. Evaluate powers.
3. Multiply and divide.
4. Add and subtract.

EXAMPLE 1   Find the value of $x + 2$ when $x = 3$.

STEP 1	Substitute 3 for x.	$x + 2 = 3 + 2$
STEP 2	Add.	$= 5$

ANSWER: **5**

EXAMPLE 2   Find the value of rt when $r = 40$ and $t = 3$.

STEP 1	Substitute 40 for r and 3 for t.	$rt = 40 \cdot 3$
STEP 2	Multiply.	$= 120$

ANSWER: **120**

EXAMPLE 3   Find the value of $3y - 7$ when $y = 5$.
(Multiply before subtracting.)

STEP 1	Substitute 5 for y.	$3y - 7 = 3(5) - 7$
STEP 2	Multiply.	$= 15 - 7$
STEP 3	Subtract.	$= 8$

ANSWER: **8**

EXAMPLE 4   Evaluate $4(x - y)$ when $x = 6$ and $y = 3$.

STEP 1	Substitute 6 for x and 3 for y.	$4(x - y) = 4(6 - 3)$
STEP 2	Evaluate within the parentheses.	$= 4(3)$
STEP 3	Multiply.	$= 12$

ANSWER: **12**

EXAMPLE 5   Evaluate $2(x + y)^2 - 4x$ when $x = 5$ and $y = -2$.

STEP 1	Substitute 5 for x and –2 for y.	$2(x + y)^2 - 4x = 2(5 + -2)^2 - 4(5)$
STEP 2	Evaluate within the parentheses.	$= 2(3)^2 - 4(5)$
STEP 3	Evaluate the power.	$= 2(9) - 4(5)$
STEP 4	Multiply.	$= 18 - 20$
STEP 5	Subtract.	$= -2$

ANSWER: **–2**

Find the value of each algebraic expression.

1. $x + 9$ when $x = -4$ $y - 12$ when $y = 3$ $z + (-5)$ when $z = 9$

2. $5n$ when $n = 4$ $7r$ when $r = -2$ $\frac{2}{3}s$ when $s = 24$

3. $\frac{x}{5}$ when $x = 35$ $\frac{y}{3}$ when $y = -18$ $\frac{a}{b}$ when $a = -10$, $b = -2$

4. xy when $x = -2$, $y = 4$ mn when $m = -3$, $n = -2$ uv when $u = 4$, $v = -2$

5. $12 + 5a$ when $a = 9$ $6x - 12$ when $x = 2$ $25 - xy$ when $x = 4$, $y = 3$

6. $ab - \frac{a}{b}$ when $a = 6$, $b = 2$ $4x + \frac{x}{3}$ when $x = -9$ $\frac{c}{d} + 3c - 2d$ when $c = 8$, $d = 4$

7. $3(x + 7)$ when $x = 2$ $-4(y - 5)$ when $y = 3$ $7(a - b)$ when $a = 2$, $b = 3$

8. $2x^2 + 3x$ when $x = 3$ $a^2 + b^2$ when $a = 6$, $b = 8$ $3(c - d)^2 + 2c$ when $c = 9$, $d = 8$

In an expression involving division, the division bar separates the problem into two parts. Evaluate the numerator and the denominator separately before dividing. Be sure to follow the order of operations shown on page 46.

EXAMPLE 6 Find the value of $\dfrac{3(y-4)^2}{2(y-1)}$ when $y = 6$.

STEP 1 Substitute 6 for y.

$$\frac{3(y-4)^2}{2(y-1)} = \frac{3(6-4)^2}{2(6-1)}$$

STEP 2 Evaluate within the parentheses.

$$= \frac{3(2)^2}{2(5)}$$

STEP 3 Evaluate the power.

$$= \frac{3(4)}{2(5)}$$

STEP 4 Multiply.

$$= \frac{12}{10}$$

STEP 5 Simplify the improper fraction.

$$= \frac{6}{5} = \mathbf{1\tfrac{1}{5}}$$

ANSWER: $1\tfrac{1}{5}$

Find the value of each algebraic expression.

9. $\dfrac{2(x+3)^2}{5(x-1)}$ when $x = 2$

$\dfrac{3(m-n)^2}{2(m+n)}$ when $m = -1, n = 3$

10. $\dfrac{4z(z-4)^2}{5z^2}$ when $z = -1$

$\dfrac{2x(x-1)^2}{3y(y+1)^2}$ when $x = 2, y = 1$

Evaluating Formulas

A **formula** is a mathematical rule involving an algebraic expression. For example, the formula for the area of a rectangle is

Area = length times width, or $A = lw$.

In a formula, the quantity that you are trying to find is usually written to the left of the equals sign. The algebraic expression is written to the right.

To find a value using a formula, substitute numbers for variables in the algebraic expression and do the indicated operations.

EXAMPLE   Use the formula $A = lw$ to find the area of a rectangle that has a length of 8 feet and a width of 5 feet.

Substitute 8 for l and 5 for w in the formula $A = lw$.
$A = 8 \times 5 = 40$

ANSWER: $A = 40$ square feet

(**Note:** In area problems, the answers are in square units. In volume problems, the answers are in cubic units.)

Find each value as indicated.

1. *Description:*   Perimeter of a rectangle
   Formula: $P = 2(l + w)$
   Variables: l = length
                     w = width

   Find P when $l = 7$ feet and $w = 4$ feet.

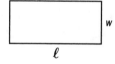

2. *Description:*   Perimeter of a triangle
   Formula: $P = a + b + c$
   Variables: a, b, c = sides of a triangle

   Find P when $a = 6$ feet, $b = 9$ feet, and $c = 13$ feet.

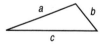

3. *Description:*   Circumference of a circle
   Formula: $C = 2\pi r$, where $\pi \approx \frac{22}{7}$
   Variables: r = radius

   Find C when $r = 7$ inches.

4. *Description:*   Area of a rectangle
    Formula: $A = lw$
    Variables: l = length
                    w = width

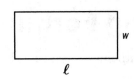

Find A when $l = 12$ feet and $w = 7$ feet.

5. *Description:*   Area of a triangle
    Formula: $A = \frac{1}{2}bh$
    Variables: b = base
                    h = height

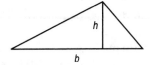

Find A when $b = 14$ inches and $h = 6$ inches.

6. *Description:*   Area of a circle
    Formula: $A = \pi r^2$, where $\pi \approx \frac{22}{7}$
    Variables: r = radius

Find A when $r = 14$ inches.

7. *Description:*   Volume of a rectangular solid
    Formula: $V = lwh$
    Variables: l = length, w = width, h = height

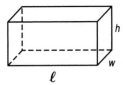

Find V when $l = 16$ inches, $w = 5$ inches,
and $h = 9$ inches.

8. *Description:*   Volume of a cylinder
    Formula: $V = \pi r^2 h$, where $\pi \approx \frac{22}{7}$
    Variables: r = radius, h = height

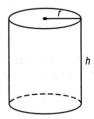

Find V when $r = 2$ feet and $h = 7$ feet.

Applying Your Skills: Formulas

Using the formulas on the previous two pages, solve the problems below.

- Use the perimeter formulas for finding the distance around a figure.
- Use the circumference formula for finding the distance around a circle.
- Use area formulas for finding the surface area of a figure.
- Use volume formulas for finding the volume of (or the space inside of) an object.

1. Working as a carpenter, you are enclosing a rectangular window with molding. How many feet of molding do you need if the window measures 5 feet long by 4 feet wide?

2. A school playground in the shape of a triangle has sides of 150 yards, 125 yards, and 191 yards. What is the distance around the playground?

3. About how many feet of fence do you need to enclose a circular garden that has a radius of 28 feet?

4. Suppose you are putting new tile on your kitchen floor. How many square feet of tile do you need if the floor measures 15 feet long by 12 feet wide?

5. A triangular roof gable is 12 yards across and 3 yards high. Find the area of this gable.

6. A field in the shape of a circle has a radius of 21 meters. Find the approximate area of this field.

7. Carlos rented a moving van that has an enclosed section measuring 18 feet by 6 feet by 7 feet. How many cubic feet of storage space does the truck have?

Simplifying an Algebraic Expression

An algebraic expression is made up of one or more **terms.** Each term is a number, a variable, or the product of a number (coefficient) and a variable.

Examples of terms are $3x$, $4n^2$, $2xy$, z, and 9. Every term involving one or more variables has a coefficient. If no coefficient is written, the coefficient is $+1$. In the examples above, the coefficient of the term z is $+1$ (usually just written 1). A number standing alone, such as 9, is called a **constant term.**

Like terms are groups of two or more terms that have the same variables raised to the same exponents. All constant terms are like terms.

Examples of like terms: $4x^2$ and $2x^2$, $3xy$ and $7xy$, a^2b and $5a^2b$, 8 and -3.

Unlike terms are groups of two or more terms in which one or more variables are either different or contain different exponents.

Examples of unlike terms: x^2 and y^2, $2n$ and n^2, a^2b and ab^2, x and 5.

Combining Terms and Simplifying Expressions

An algebraic expression often contains more than one term. To **simplify an algebraic expression,** you combine like terms. You cannot combine unlike terms.

To combine like terms, combine the coefficients of the like terms. The sign $(+$ or $-)$ to the left of each coefficient stays with the coefficient.

Algebraic Expression	Simplified Expression
$5x + 3x$	$8x$
$7y - 4y + 5$	$3y + 5$
$2z + z + 3 + 6$	$3z + 9$
$3x^2 + 7x + 2x^2 - 3x$	$5x^2 + 4x$

Parentheses in expressions must be removed before terms can be combined. To remove parentheses, multiply each term within the parentheses by the coefficient.

EXAMPLES $\quad 4(x^2 + 2x) - x^2 = 4x^2 + 8x - x^2 = (4 - 1)x^2 + 8x = 3x^2 + 8x$

$\qquad\qquad 3(y^3 - 2y^2) + 2y^3 = 3y^3 - 6y^2 + 2y^3 = (3 + 2)y^3 - 6y^2 = 5y^3 - 6y^2$

Identify the coefficient part (C) and the variable part (V) of each term.
The first problem in each row is done as an example

	C	V		C	V		C	V
1. $-4x$	-4	x	$3a$			$-5z$		
2. y^2	1	y^2	z^2			$-a^2$		
3. $-9x^2y$	-9	x^2y	$-4c^2d$			$7uv^2$		
4. $3a^2b^2$	3	a^2b^2	$8x^2y^2$			$-7ab^3$		

Circle the pair of like terms in each group.

5. $x, x^2, -5x$　　　　　　　$-y^3, 3y, -2y^3$　　　　　　$a^2b, 2a^2b, b^2a$

6. $a^2, -4a^2, 2a, 5a^4$　　　　$-2x, 2xy, -2x^2y, xy^2$　　　$-3x, -3x^2, x^3, 4x^3$

7. $2c, -c^2, c^3, -5c^3$　　　　$3rs, -rs^2, s^2, 3rs^2$　　　$-4x^3y, xy^4, 3x^3y^3, 2x^3y$

Write a simplified algebraic expression for each perimeter below.

8. $P =$ _____

9. $P =$ _____

10. $P =$ _____

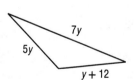

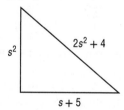

11. $P =$ _____

12. $P =$ _____

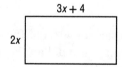

Algebraic Expressions Review

Solve the problems below. When you finish, check your answers at the back of the book. Then correct any errors.

Write an algebraic expression for each word expression.

1. 7 subtracted from x _____

2. z divided by 9 _____

3. 19 added to y _____

4. -4 times n _____

5. 6 times the quantity x plus 4 _____

6. 2 times the square of the quantity $y - 5$ _____

Write a word expression for each algebraic expression.

7. $12 + y$ _____

8. $x - 9$ _____

9. $15a$ _____

10. $\dfrac{y}{12}$ _____

11. $4s + 8$ _____

12. $-9(2x + 4)$ _____

Write an algebraic expression as indicated.

13. A notebook is on sale for s dollars. Write an algebraic expression for the cost of six notebooks at this price.

14. Arlene paid m dollars for groceries that cost $12.49. Write an algebraic expression that tells the amount of change Arlene should receive.

15. Janine took a ribbon that was n feet long and cut off a 2-foot piece. She then divided the remaining ribbon into six smaller pieces. Write an algebraic expression that tells the length of each smaller piece.

16. A pair of shoes is on sale for 40% off. The regular price of the shoes is n dollars. Write an algebraic expression that tells the sale price of the shoes.

Find the value of each algebraic expression.

17. $13 + 5x$ when $x = -3$

18. $7y - 8$ when $y = 6$

19. $\frac{z}{3} + 9$ when $z = -6$

20. $4xy$ when $x = 5, y = 3$

21. $-3(z + 7)$ when $z = -4$

22. $2a^2 - 4ab$ when $a = 2$, $b = 3$

23. $(m + n)(m - n)$ when $m = -4, n = 2$

24. $\dfrac{2(x + y)^2}{4(x - y)}$ when $x = 3, y = 1$

Solve.

25. Use the formula $V = \pi r^2 h$, where $\pi \approx \frac{22}{7}$, to estimate V when $r = 3$ feet and $h = 10$ feet. Round to the nearest whole number.

26. The formula $i = prt$ is used to calculate simple interest in dollars, where p is the principal, r is the interest rate, and t is the time in years. Find i when $p = \$550$, $r = 0.09$, and $t = 1.5$.

27. Suppose your kitchen is in the shape of a rectangle that is 15 feet long by 12 feet wide. Use the area formula $A = lw$ to find how many square feet of tile you will need to tile the kitchen floor.

28. Use the temperature formula $°C = \frac{5}{9}(°F - 32)$ to find the Celsius temperature ($°C$) when the Fahrenheit temperature ($°F$) is $104°F$.

ONE-STEP EQUATIONS

What Is an Equation?

An **equation** is a statement that identifies two quantities as equal. You may see an equation written in words or in mathematical symbols.

Equation in words: Two plus three is equal to five.
Equation in symbols: $2 + 3 = 5$

(**Note**: The equals sign (=) is read "is equal to" or simply "equals.")

Equations are common in mathematics. In daily life, mathematical questions often start as word problems and are solved as equations. We put words into symbols because symbols are easier to work with. Each time you add, subtract, multiply, or divide, you use an equation.

Numeric
Equations
$\begin{cases} \text{Addition Equation:} & 32 + 19 = 51 \\ \text{Subtraction Equation:} & 76 - 30 = 46 \\ \text{Multiplication Equation:} & 43 \times 17 = 731 \\ \text{Division Equation:} & \frac{80}{5} = 16 \end{cases}$

In algebra, an equation may contain one or more letters in place of numbers. Each letter stands for a number whose value may not yet be known.

Algebraic
Equations
$\begin{cases} \text{Addition Equation:} & x + 23 = 47 \\ \text{Subtraction Equation:} & y - 52 = 15 \\ \text{Multiplication Equation:} & 9a = 72 \\ \text{Division Equation:} & \frac{b}{5} = 16 \end{cases}$

Following are examples of how to read algebraic equations.

$x + 23 = 47$ is read "x plus 23 is equal to 47"
$y - 52 = 15$ is read "y minus 52 equals 15"
$9a = 72$ is read "9 times a equals 72"
$\frac{b}{5} = 16$ is read "b divided by 5 is equal to 16"

> There is more than one correct way to read an algebraic equation.

Circle the equation described by the words.

1. The product of nine and a number is equal to fourteen.

   a. $n + 9 = 14$ **b.** $9n = 14$ **c.** $14n = 9$ **d.** $\frac{9}{n} = 14$

2. The quotient of a number divided by six equals fifty.

   a. $6n = 50$ **b.** $\frac{6}{n} = 14$ **c.** $\frac{n}{6} = 50$ **d.** $6 \times 50 = n$

Write an addition equation for each drawing.

3.

Equation:

4.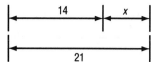

Equation:

Write a subtraction equation for each drawing.

5.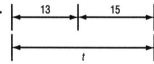

Equation:

6.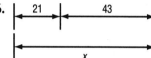

Equation:

Write a multiplication equation for each drawing.

7.

Equation:

8.

Equation:

Write a division equation for each drawing.

9.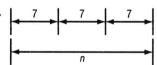

Equation:

10.

Equation:

Checking Unknowns

As you have seen, an algebraic equation contains a letter (variable) in place of a number. In an equation, this letter is often called an **unknown** because its value is unknown.

Solving an equation means finding the value for the variable that makes the equation true. The **solution** is the value of the unknown that solves the equation.

To check if a value of the unknown is the correct solution, follow these two steps.

STEP 1 Substitute the value for the unknown into the equation.

STEP 2 Do the operations and compare each side of the equation.

EXAMPLE 1 Is $y = 5$ the solution for $3y - 9 = 6$?

STEP 1 Substitute 5 for y. $3(5) - 9 = 6$

STEP 2 Subtract. Compare. $15 - 9 = 6$
                                                          $6 = 6$

Since $6 = 6$, **$y = 5$ is the solution** to the equation.

EXAMPLE 2 Is $x = 8$ the solution for $x + 7 = 14$?

STEP 1 Substitute 8 for x. $8 + 7 \neq 14$

STEP 2 Add. Compare. $15 \neq 14$

> The symbol $\neq$ means "is not equal to."

Since 15 is not equal to 14, **$x = 8$ is not the solution** to the equation.

Determine if the suggested value is the solution to the equation. Circle Yes if the value is the solution, No if it is not.

1. $x + 3 = 9$ Try $x = 4$.   Yes   No

2. $y - 12 = 37$ Try $y = 49$.   Yes   No

3. $3z = 39$ Try $z = 13$.   Yes   No

4. $\frac{a}{15} = 3$ Try $a = 54$.   Yes   No

5. $\frac{3}{4}m = 3$ Try $m = 4$.   Yes   No

6. $n + (-4) = 15$ Try $n = 11$.   Yes   No

7. $x - (-3) = 10$ Try $x = 7$.   Yes   No

8. $4s = -20$ Try $s = -5$.   Yes   No

Solving an Algebraic Equation

An algebraic equation is like a balance. The left side of the equation balances, or equals, the right side of the equation.

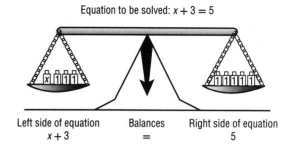

Equation to be solved: $x + 3 = 5$

Left side of equation — Balances — Right side of equation
$x + 3$ — = — 5

The addition equation $x + 3 = 5$ can be represented as weights on a balance. On the left side are the unknown weight x and 3 unit weights. On the right side are 5 unit weights.

To find x, *remove an equal amount from each side* until x stands alone. Removing an equal amount from each side ensures that the equation remains balanced.

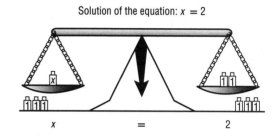

Solution of the equation: $x = 2$

x — = — 2

You have solved the equation when x stands alone on the left and the value that balances x stands alone on the right.

The equation $x + 3 = 5$ is solved by *subtracting 3 from each side.* As shown at the right, $x = 2$.

Subtraction is used to solve the addition equation $x + 3 = 5$. Because subtraction *undoes* addition, subtraction is called the **inverse** (opposite) of addition.

As you will see in the pages ahead, all algebraic equations are solved by using **inverse operations.** An inverse operation *undoes* an operation and leaves the unknown standing alone.

Operation		Inverse Operation	Solving an Equation
addition	⟷	subtraction	Use subtraction to solve an addition equation.
subtraction	⟷	addition	Use addition to solve a subtraction equation.
multiplication	⟷	division	Use division to solve a multiplication equation.
division	⟷	multiplication	Use multiplication to solve a division equation.

The example $x + 3 = 5$ also points out the most important rule to remember as you use inverse operations to solve equations:

Rule: Any inverse operation performed on one side of the equation must also be performed on the other side of the equation.

Solving an Addition Equation

To solve an addition equation, subtract the *added number* from each side of the equation. The variable will be left alone, and the equation will be solved.

EXAMPLE   Solve for x: $x + 6 = 15$ **Solve:** $x + 6 = 15$

       STEP 1 Subtract 6 from each side of the equation. $x + 6 - 6 = 15 - 6$

       STEP 2 Simplify both sides of the equation. $x = 9$
                  On the left, $6 - 6 = 0$, leaving x alone.
                  On the right, $15 - 6 = 9$.

ANSWER: $x = 9$

Check the answer by substituting 9 for x.   **Check:** $9 + 6 = 15$
                                                  $\checkmark 15 = 15$

Solve each addition equation and check each answer. The first problem in each row is partially completed. Show all steps.

1. $x + 5 = 8$ $x + 7 = 12$ $y + 3 = 9$ $n + 10 = 24$
$x + 5 - 5 = 8 - 5$

2. $y + 2 = 15$ $s + 6 = 11$ $x + 5 = 28$ $z + 17 = 35$
$y + 2 - 2 = 15 - 2$

3. $x + \frac{1}{2} = 4$ $y + \frac{2}{3} = 3$ $c + \frac{3}{4} = \frac{5}{4}$ $r + 1\frac{5}{8} = 5\frac{7}{8}$

$x + \frac{1}{2} - \frac{1}{2} = 4 - \frac{1}{2}$

4. $x + 3.2 = 6.8$ $n + 2.4 = 5.7$ $z + 5.5 = 14$ $y + 2.75 = 7.5$
$x + 3.2 - 3.2 = 6.8 - 3.2$

Solving a Subtraction Equation

To solve a subtraction equation, add the *subtracted number* to each side of the equation. The variable will be left alone, and the equation will be solved.

EXAMPLE   Solve for x: $x - 7 = 12$ **Solve:** $x - 7 = 12$

STEP 1   Add 7 to each side of the equation. $x - 7 + 7 = 12 + 7$

STEP 2   Simplify both sides of the equation. $x = 19$
On the left, $-7 + 7 = 0$, leaving x alone.
On the right, $12 + 7 = 19$.

ANSWER: $x = 19$

Check the answer by substituting 19 for x.   **Check:** $19 - 7 = 12$
✓ $12 = 12$

Solve each subtraction equation and check each answer. The first problem in each row is partially completed. Show all steps.

1. $x - 6 = 14$ $x - 3 = 7$ $y - 6 = 10$ $z - 15 = 32$
$x - 6 + 6 = 14 + 6$

2. $y - 5 = 12$ $x - 8 = 18$ $y - 11 = 26$ $n - 9 = 27$
$y - 5 + 5 = 12 + 5$

3. $x - \frac{2}{3} = 6$ $r - \frac{5}{8} = 3$ $y - \frac{1}{4} = \frac{7}{4}$ $v - 2\frac{1}{2} = 5\frac{1}{2}$
$x - \frac{2}{3} + \frac{2}{3} = 6 + \frac{2}{3}$

4. $m - 2.4 = 5.3$ $x - 1.7 = 6.4$ $n - 7.5 = 9.75$ $y - 3.25 = 8.5$
$m - 2.4 + 2.4 = 5.3 + 2.4$

Applying Your Skills: One-Step Equations (+ and −)

Algebraic equations can be written for problems that can easily be solved without algebra. Practicing algebra skills here, though, is an excellent way to become more confident with the use of equations.

Circle the correct equation. Then find the value of the variable.

1. If you subtract 18 from a number, the difference is 14. If n stands for the unknown number, which equation can be used to find n?

   a. $18 - n = 14$

   b. $n - 14 = 18$

   c. $n - 18 = 14$

   d. $18 - 14 = n$

   $n =$

2. When you started work today, you placed $125.00 in the cash register. At the end of the day, the cash register contains $435.75. Which equation can be used to find the total t of today's sales?

   a. $t - \$435.75 = \$125.00$

   b. $t - \$125.00 = \$435.75$

   c. $t + \$125.00 = \$435.75$

   d. $t + \$435.75 = \$125.00$

   $t =$

3. Lynn bought a pair of shoes that were on sale for 25% off the regular price of $40. Which equation can be used to find the amount p Lynn paid?

   a. $p + \$40 = 0.25(\$40)$

   b. $p - 0.25(\$40) = \$40$

   c. $p - \$40 = 0.25(\$40)$

   d. $p + 0.25(\$40) = \$40$

   $p =$

4. The temperature in Chicago rose 15°F between 9:00 A.M. and noon. If the temperature at noon is 87°F, which equation can be used to find the temperature t at 9:00 A.M.?

   a. $t - 15°F = 87°F$

   b. $t - 87°F = 15°F$

   c. $t + 15°F = 87°F$

   d. $t = 15°F + 87°F$

   $t =$

Write an equation for each problem. Then solve your equation to find the value of the variable.

5. Ayla bought a sweater on sale for $34.00, $14.95 less than its original price. Write an equation to find the original price p of the sweater.

   Equation:

   $p =$

6. Blake had a full bag of cookies. After giving 14 cookies away, he still had 19 left. Write an equation to find the number n of cookies that were in the bag originally.

   Equation:

   $n =$

Solving a Multiplication Equation

To solve a multiplication equation, divide each side of the equation by the number that multiplies the variable. The variable will be left alone, and the equation will be solved.

EXAMPLE Solve for x: $4x = 32$ **Solve:** $4x = 32$

STEP 1 Divide each side of the equation by 4.

$$\frac{^1\cancel{4}x}{\cancel{4}_1} = \frac{32}{4}$$

STEP 2 Simplify each side of the equation. On the left, $\frac{4}{4} = 1$, leaving x alone. On the right, $\frac{32}{4} = 8$.

$$x = 8$$

ANSWER: $x = 8$

Check the answer by substituting 8 for x. **Check:** $4(8) = 32$

$\checkmark 32 = 32$

Solve each multiplication equation and check each answer. The first problem in each row is partially completed. Show all steps.

1. $7x = 35$ $8n = 48$ $6y = 54$ $7m = 49$

$$\frac{7x}{7} = \frac{35}{7}$$

2. $-4y = 24$ $-3x = 27$ $-5b = 65$ $-9y = 99$

$$\frac{-4y}{-4} = \frac{24}{-4}$$

3. $6x = 18$ $8n = 56$ $7z = 147$ $11y = 121$

$$\frac{6x}{6} = \frac{18}{6}$$

4. $7y = 16.8$ $5y = 32.5$ $4.2x = 22.68$ $8.5n = 63.75$

$$\frac{7y}{7} = \frac{16.8}{7}$$

Solving a Division Equation

To solve a division equation, multiply each side of the equation by the number that divides the variable. The variable will be left alone, and the equation will be solved.

EXAMPLE   Solve for x: $\frac{x}{6} = 14$ **Solve:** $\frac{x}{6} = 14$

STEP 1   Multiply each side of the equation by 6. $\frac{x}{6}(6) = 14(6)$

STEP 2   Simplify each side of the equation. $x = 84$
On the left, $\frac{6}{6} = 1$, leaving x alone.
On the right, $14(6) = 84$.

ANSWER: $x = 84$

Check the answer by substituting 84 for x.   **Check:** $\frac{84}{6} = 14$
$\checkmark\, 14 = 14$

Solve each division equation and check each answer. The first problem in each row is partially completed. Show all steps.

1. $\frac{x}{5} = 4$ $\frac{y}{3} = 9$ $\frac{c}{8} = 11$ $\frac{n}{4} = 12$

$\frac{x}{5}(5) = 4(5)$

2. $\frac{y}{3} = 6$ $\frac{x}{4} = 17$ $\frac{n}{3} = 12$ $\frac{t}{12} = 15$

$\frac{y}{3}(3) = 6(3)$

3. $\frac{x}{2} = 3\frac{1}{2}$ $\frac{y}{3} = 2\frac{1}{3}$ $\frac{n}{6} = 2.5$ $\frac{r}{4} = 3.375$

$\frac{x}{2}(2) = 3\frac{1}{2}(2)$

Applying Your Skills: One-Step Equations ($\times$ and $\div$)

On this page, you will practice using algebraic equations to solve multiplication and division problems.

Circle the correct equation. Then find the value of the variable.

1. If you multiply a number n by 14, the product is 168. Which equation below can be used to find n?

   a. $168n = 14$

   b. $\frac{14}{n} = 168$

   c. $14n = 168$

   d. $n = 168(14)$

   $n =$

2. On his diet, Brett lost 5% of his body weight. If Brett lost 9 pounds, which equation can be used to find his body weight w before he started dieting?

   a. $\frac{w}{0.05} = 9$

   b. $0.05w = 9$

   c. $9w = 0.05$

   d. $w = 9(0.05)$

   $w =$

3. One out of four students in Marie's pottery class are men. If 7 students in the class are men, which equation can be used to find the number n of students in the class?

   a. $\frac{n}{4} = 7$

   b. $7n = \frac{1}{4}$

   c. $\frac{n}{7} = \frac{1}{4}$

   d. $4n = 7$

   $n =$

4. As an office assistant you decide to buy 3 reams of printer paper. The total cost is $16.50. Which equation can be used to find the average price p of a single ream?

   a. $\frac{p}{3} = \$16.50$

   b. $p = \frac{\$16.50}{3}$

   c. $p = \frac{3}{\$16.50}$

   d. $p(\$16.50) = 3$

   $p =$

Write an equation for each problem. Then solve your equation to find the variable.

5. For every hour of overtime you work on weekends, your boss gives you 1.5 hours of paid time off during slower times. Last month you earned 42 hours of paid time off. How many hours h of overtime did you work last month?

   Equation:

   $h =$

6. One out of eight of the students in Ms. Knapp's kindergarten class are sick with the flu. If 4 students are sick, how many students s are in Ms. Knapp's class?

   Equation:

   $s =$

Solving an Equation with a Fraction Coefficient

To solve an equation with a **fraction coefficient,** multiply each side of the equation by the **reciprocal** of the fraction. The variable will be left alone, and the equation will be solved. The reciprocal is found by interchanging the numerator and the denominator of a fraction.

| EXAMPLE | Solve for x: $\frac{2}{5}x = 16$ | **Solve:** $\frac{2}{5}x = 16$ |

STEP 1 Multiply each side of the equation by $\frac{5}{2}$.

$$\frac{5}{2}\left(\frac{2}{5}\right)x = \frac{5}{2}(16)$$

STEP 2 Simplify each side of the equation.

$$x = 40$$

On the left, $\frac{5}{2}\left(\frac{2}{5}\right) = 1$, leaving x alone.

On the right, $\frac{5}{2}(16) = 40$.

ANSWER: $x = 40$

Check the answer by substituting 40 for x.

Check: $\frac{2}{5}(40) = 16$
✓ $16 = 16$

Solve each equation and check each answer. The first problem in each row is partially completed. Show all steps.

1. $\frac{5}{8}x = 5$ $\frac{5}{6}z = 10$ $\frac{3}{7}a = 12$ $\frac{2}{3}n = 8$

$$\frac{8}{5}\left(\frac{5}{8}\right)x = \frac{8}{5}(5)$$

2. $-\frac{3}{7}n = +\frac{3}{4}$ $-\frac{1}{2}s = \frac{1}{2}$ $-\frac{3}{4}x = \frac{2}{3}$ $-\frac{2}{5}y = \frac{7}{8}$

$$-\frac{7}{3}\left(-\frac{3}{7}\right)n = -\frac{7}{3}\left(\frac{3}{4}\right)$$

3. $\frac{2}{3}z = 3.50$ $\frac{3}{4}x = 1.20$ $\frac{3}{8}c = 6.60$ $\frac{5}{6}r = 1.75$

$$\frac{3}{2}\left(\frac{2}{3}\right)z = \frac{3}{2}(3.50)$$

Applying Your Skills: One-Step Equations (with Fractions)

On this page, you will practice using algebraic equations that involve fraction coefficients.

Circle the correct equation. Then find the value of the variable.

1. If you multiply a number x by $\frac{2}{3}$, the product is 36. Which equation below can be used to find x?

   a. $x \div \frac{2}{3} = 36$

   b. $\frac{2}{3}x = 36$

   c. $36x = \frac{2}{3}$

   d. $x = \left(\frac{2}{3}\right)36$

   $x =$

2. Suppose $\frac{1}{3}$ of your time in class is spent studying. If you spent 4 hours studying in class last week, which equation tells how many total hours h you spent in class?

   a. $h = 4 \div \frac{1}{3}$

   b. $\frac{2}{3}h = 4$

   c. $h = \frac{1}{3}(4)$

   d. $\frac{1}{3}h = 4$

   $h =$

3. One-fourth of the students in Lysha's class speak Spanish. If the class contains 18 students who do *not* speak Spanish, which equation can be used to find the total number n of students in the class?

   a. $\frac{1}{4}n = 18$

   b. $\frac{2}{4}n = 18$

   c. $\frac{3}{4}n = 18$

   d. $n = \left(\frac{3}{4}\right)18$

   $n =$

4. At a sale, Yvette paid $7.75 for an umbrella labeled 20% off. Which equation can be used to find the regular price r of the umbrella?

   a. $\frac{1}{5}r = \$7.75$

   b. $\frac{4}{5}r = \$7.75$

   c. $\frac{6}{5}r = \$7.75$

   d. $r = \left(\frac{1}{5}\right)\$7.75$

   $r =$

Write an equation for each problem. Then solve your equation to find the variable.

5. You find that $\frac{2}{5}$ of your time as an office assistant is spent on the phone with clients. If client phone calls take 14 hours each week, how many total hours h do you work each week?

   Equation:

   $h =$

6. As a salesman at Valley Auto, five-eighths of the cars you sell are foreign-made. If you sold 50 foreign-made cars last year, how many total cars c did you sell?

   Equation:

   $c =$

One-Step Equations Review

Solve the problems below. When you finish, check your answers at the back of the book. Then correct any errors.

Solve each addition equation by subtraction.

1. $x + 8 = 10$

2. $z + (-5) = 8$

3. $a + 5 = 2$

Solve each subtraction equation by addition.

4. $y - 7 = 12$

5. $x - 12 = 13$

6. $b - 3 = -7$

Solve each multiplication equation by division.

7. $9z = 72$

8. $4x = -32$

9. $-13y = 143$

Solve each division equation by multiplication.

10. $\frac{x}{3} = 9$

11. $\frac{y}{7} = 8$

12. $\frac{a}{6} = -4$

Solve each equation with a fraction coefficient by multiplying by the reciprocal of the coefficient.

13. $-\frac{2}{5}x = 6$

14. $\frac{4}{9}y = 16$

15. $\frac{7}{4}a = -14$

Solve each equation.

16. $y - 12 = 5$

17. $\frac{4}{7}n = 20$

18. $5q = 120$

19. $\frac{y}{5} = 7$

20. $4n = -28$

21. $x + 10 = 7$

22. $6p = 42$

23. $x + (-7) = 18$

24. $s - 9 = -9$

25. $x - \frac{1}{6} = \frac{1}{2}$

26. $\frac{r}{3} = 2.25$

27. $5x = \frac{1}{3}$

28. $3k = 9.36$

29. $\frac{x}{3} = \frac{2}{5}$

30. $n - \frac{2}{3} = 1\frac{1}{3}$

31. $\frac{5}{6}z = 35$

32. $y - 3.25 = 8$

33. $x + \frac{3}{4} = 5\frac{1}{4}$

For each problem,

    a. **choose a variable to represent the unknown.**
    b. **write an equation using the variable.**
    c. **solve the equation.**

The first problem is done as an example.

34. Zeta spent $4.75 for breakfast and had $2.40 left over. How much money did Zeta have before paying for breakfast?

    a. Unknown: m = money Zeta had

    b. Equation: $m - \$4.75 = \$2.40$

    c. Solution:

$$m - \$4.75 + \$4.75 = \$2.40 + \$4.75$$
$$\boldsymbol{m = \$7.15}$$

$$\begin{array}{r} \overset{1}{\$4.75} \\ +\ \ 2.40 \\ \hline \$7.15 \end{array}$$

35. Heather divided a box of cookies among five people. If each person received six cookies, how many cookies were in the box?

    a. Unknown:

    b. Equation:

    c. Solution:

36. You are buying supplies for the restaurant where you work. You receive $5.23 in change. If the supplies cost $124.77, how much did you give the clerk?

    a. Unknown:

    b. Equation:

    c. Solution:

37. Three-fourths of the students in Jenni's gymnastics class are girls. If there are 18 girls in the class, how many students are in the class?

    a. Unknown:

    b. Equation:

    c. Solution:

38. One out of six teachers at Grant School drives a foreign-made car. If four teachers drive a foreign-made car, how many teachers does Grant School have?

    a. Unknown:

    b. Equation:

    c. Solution:

39. Jocelyn saves $\frac{1}{3}$ of the money she earns babysitting and spends the rest. If she spent $90 during May, how much did Jocelyn earn babysitting during May?

    a. Unknown:

    b. Equation:

    c. Solution:

Write an equation for each drawing. Solve your equation. (Note: More than one equation is possible, but only one solution is correct.)

40.

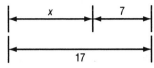

Equation:

$x =$

41.

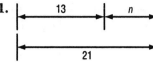

Equation:

$n =$

42.

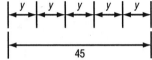

Equation:

$y =$

43.

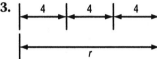

Equation:

$r =$

Write an equation for the perimeter of each triangle and solve for the length of the missing side. (Remember: $P = s_1 + s_2 + s_3$) Problem 44 is started for you.

44.

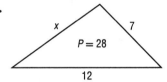

Equation: $x + 7 + 12 = 28$
$\qquad\qquad x + 19 = 28$

$x =$

45.

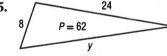

Equation:

$y =$

Write an equation for the area of each rectangle and solve for the length of the missing side. (Remember: $A = lw$) Problem 46 is started for you.

46.

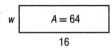

Equation: $16w = 64$

$w =$

47.

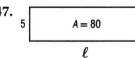

Equation:

$l =$

MULTISTEP EQUATIONS

What Is a Multistep Equation?

A **multistep equation** is an equation that contains more than one operation. To solve a multistep equation, follow these two rules.

> **Rule 1:** Do addition or subtraction first.
>
> a. If a number is added to the variable, subtract that number from each side of the equation.
>
> b. If a number is subtracted from a variable, add that number to each side of the equation.

> **Rule 2:** Do multiplication or division last.
>
> a. If the variable is multiplied by a number, divide each side of the equation by that number.
>
> b. If the variable is divided by a number, multiply each side of the equation by that number.

EXAMPLE 1   Solve for x: $2x - 7 = 19$

Solve: $2x - 7 = 19$

    STEP 1   Add 7 to each side of the equation.

$$2x - 7 + 7 = 19 + 7$$
$$2x = 26$$

    STEP 2   Divide each side by 2.

$$\frac{2}{2}x = \frac{26}{2}$$
$$x = 13$$

ANSWER: $x = 13$

Check by substituting 13 for x.

Check:
$$2(13) - 7 = 19$$
$$26 - 7 = 19$$
$$\checkmark\ 19 = 19$$

EXAMPLE 2   Solve for z: $\frac{z}{4} + 6 = 22$

Solve: $\frac{z}{4} + 6 = 22$

    STEP 1   Subtract 6 from each side.

$$\frac{z}{4} + 6 - 6 = 22 - 6$$
$$\frac{z}{4} = 16$$

    STEP 2   Multiply each side by 4.

$$4\left(\frac{z}{4}\right) = 4(16)$$
$$z = 64$$

ANSWER: $z = 64$

Check by substituting 64 for z.

Check:
$$\frac{64}{4} + 6 = 22$$
$$16 + 6 = 22$$
$$\checkmark\ 22 = 22$$

Solve each equation and check your answer. The first problem in each row is done as an example.

1. $5y + 4 = 19$ $3x + 11 = 20$ $2z + 9 = 1$

$$5y + 4 - 4 = 19 - 4$$
$$5y = 15$$
$$\frac{5y}{5} = \frac{15}{5}$$
$$y = 3$$

$$\checkmark 5(3) + 4 = 19$$
$$15 + 4 = 19$$
$$19 = 19$$

2. $3x - 9 = 3$ $4z - 8 = 24$ $-3a - 4 = 17$

$$3x - 9 + 9 = 3 + 9$$
$$3x = 12$$
$$\frac{3x}{3} = \frac{12}{3}$$
$$x = 4$$

$$\checkmark 3(4) - 9 = 3$$
$$12 - 9 = 3$$
$$3 = 3$$

3. $\frac{y}{4} + 5 = 3$ $\frac{z}{2} + 6 = 7$ $-\frac{x}{5} + 3 = 9$

$$\frac{y}{5} + 5 - 5 = 3 - 5$$
$$\frac{y}{4} = -2$$
$$4\left(\frac{y}{4}\right) = 4(-2)$$
$$y = -8$$

$$\checkmark -\frac{8}{4} + 5 = 3$$
$$-2 + 5 = 3$$
$$3 = 3$$

Learning a Shortcut to Solving Equations

To solve an equation, you must get the variable standing alone. In the examples below, you will learn a shortcut in which you do not write

- numbers that add to 0, or
- coefficients that equal 1.

	Shortcut	**Long Way**

Shortcut

EXAMPLE 1 **Solve:** $3x + 9 = 30$

STEP 1 Move 9 to the right side of the equation and change the + sign to a − sign.

$3x = 30 - 9$
$3x = 21$

STEP 2 Divide 21 by 3.

$x = \frac{21}{3}$

ANSWER: $x = 7$

EXAMPLE 2 **Solve:** $\frac{y}{2} - 12 = 31$

STEP 1 Move 12 to the right side of the equation and change the − sign to a + sign.

$\frac{y}{2} = 31 + 12$
$\frac{y}{2} = 43$

STEP 2 Multiply 43 by 2.
$y = 2(43)$

ANSWER: $y = 86$

Long Way

Solve: $3x + 9 = 30$

STEP 1 Subtract 9 from each side.

$3x + 9 - 9 = 30 - 9$
$3x = 21$

STEP 2 Divide each side by 3.
$\frac{3x}{3} = \frac{21}{3}$

ANSWER: $x = 7$

Solve: $\frac{y}{2} - 12 = 31$

STEP 1 Add 12 to each side.
$\frac{y}{2} - 12 + 12 = 31 + 12$
$\frac{y}{2} = 43$

STEP 2 Multiply each side by 2.
$2\left(\frac{y}{2}\right) = 2(43)$

ANSWER: $y = 86$

The steps of the shortcut can be summed up as follows.

STEP 1 If a number is added to or subtracted from a variable, move the number to the other side of the equation and change the sign.

STEP 2 If a number multiplies a variable, solve for the variable by dividing the other side of the equation by the number.

STEP 3 If a number divides a variable, solve for the variable by multiplying the other side of the equation by the number.

Use the shortcut for the rest of the equations in this chapter.

Use the shortcut to solve each equation.

1. $3x + 11 = 29$
$$3x = 29 - 11$$
$$3x = 18$$
$$x = \frac{18}{3}$$
$$x = 6$$

$4y + 9 = 37$

$7z - 6 = 29$

2. $-12 + 2y = 14$
$$2y = 14 + 12$$
$$2y = 26$$
$$y = \frac{26}{2}$$
$$y = 13$$

$-5 + 7a = 30$

$13 - 9x = -41$

3. $\frac{y}{5} + 3 = 6$
$$\frac{y}{5} = 6 - 3$$
$$\frac{y}{5} = 3$$
$$y = 5(3)$$
$$y = 15$$

$\frac{z}{2} + 6 = 4$

$-\frac{n}{3} + 4 = 12$

4. $2x + 5 = 7$

$2 + 3a = 11$

$3z - 7 = 8$

5. $-14 + 2x = 4$

$\frac{z}{3} - 2 = 5$

$\frac{3}{4}n + 3 = 9$

Solving an Equation with Separated Variables

An equation is made up of **terms.** Each term is a number standing alone or a variable multiplied by a coefficient.

When a variable appears in more than one term, the separate terms can be combined. To combine terms, add or subtract the coefficients as indicated.

For example, the terms below are combined as follows.

$$5x + 2x \rightarrow 7x \qquad \text{since } 5 + 2 = 7$$
$$7y - 4y \rightarrow 3y \qquad \text{since } 7 - 4 = 3$$
$$8z + z \rightarrow 9z \qquad \text{since } 8 + 1 = 9$$
$$6n - n \rightarrow 5n \qquad \text{since } 6 - 1 = 5$$
$$9w + (-5w) \rightarrow 4w \qquad \text{since } 9 + (-5) = 4$$

(**Note:** When no number multiplies (is in front of) a variable, the coefficient is 1.)

To solve an equation in which the variable appears in more than one term, combine the separate terms as your first step.

| EXAMPLE | Solve for x: $4x - x + 2 = 17$ | **Solve:** $4x - x + 2 = 17$ |

STEP 1 Combine the x terms. $(4 - 1 = 3)$ $\qquad\qquad 3x + 2 = 17$

STEP 2 Subtract 2 from 17. $\qquad\qquad\qquad\qquad 3x = 17 - 2$
$$3x = 15$$

STEP 3 Divide 15 by 3. $\qquad\qquad\qquad\qquad\qquad x = \dfrac{15}{3}$
$$\boldsymbol{x = 5}$$

ANSWER: $\boldsymbol{x = 5}$

Check by substituting 5 for x. $\qquad\qquad$ **Check:** $4(5) - 5 + 2 = 17$
$$20 - 5 + 2 = 17$$
$$\checkmark\ 17 = 17$$

Solve each equation.

1. $2x + 3x = 25$ $\qquad\qquad\qquad 3y + 5y = 32 \qquad\qquad\qquad 5z + z = 24$
$$5x = 25$$
$$x = \frac{25}{5}$$
$$x = 5$$

2. $3a - a = 18 + 6$

$2a = 24$

$a = \dfrac{24}{2}$

$a = 12$

$7y - 3y = 23 + 9$

$13x - 8x = 17 + 8$

3. $2x + 4x - 9 = 15$

$6x - 9 = 15$

$6x = 15 + 9$

$6x = 24$

$x = \dfrac{24}{6}$

$x = 4$

$9z - z - 5 = 11$

$12y - 11y - 9 = 17$

4. $\dfrac{3}{4}y - \dfrac{1}{2}y - 5 = 7$

$\left(\dfrac{3}{4} - \dfrac{1}{2}\right)y - 5 = 7$

$\dfrac{1}{4}y = 7 + 5$

$\text{or } \dfrac{y}{4} = 12$

$y = 4(12)$

$y = 48$

$\dfrac{3}{5}a - \dfrac{1}{5}a - 3 = 17$

$\dfrac{2}{3}x + \dfrac{1}{6}x - 4 = 11$

5. $5z + 7z = 48$

$5y - 3y = 14$

$4a + 3a - 4 = 10$

6. $4a - 2a = 7 + 13$

$\dfrac{1}{2}x + \dfrac{1}{4}x = 9$

$2y - y + 2 = 7$

7. $5b - 2b = 14 - 5$

$3z + 6z + 7 = 29 - 4$

$y - \dfrac{1}{2}y = 3$

Terms on Both Sides of an Equation

When variables are on both sides of an equation, the first step is to move them to the same side, usually the left side. To move a variable from one side of an equation to the other, simply move it and change the sign that precedes it.

EXAMPLE 1 Solve for x: $3x = 2x + 7$

STEP 1 Move $2x$ to the left side of the equation, changing the sign from $+$ to $-$.
(Subtract $2x$ from each side.)

STEP 2 Subtract $2x$ from $3x$.

ANSWER: $x = 7$

Check by substituting 7 for x.

Solve: $3x = 2x + 7$

$$3x - 2x = 7$$

$$x = 7$$

Check: $3(7) = 2(7) + 7$
$21 = 14 + 7$
$\checkmark 21 = 21$

EXAMPLE 2 Solve for y: $2y - 6 = -3y + 24$

STEP 1 Add $3y$ to the left side of the equation, changing the sign from $-$ to $+$.

STEP 2 Combine the y terms.

STEP 3 Add 6 to 24.

STEP 4 Divide 30 by 5.

Solve: $2y - 6 = -3y + 24$

$$2y + 3y - 6 = 24$$

$$5y - 6 = 24$$

$$5y = 24 + 6$$
$$5y = 30$$

$$y = \frac{30}{5}$$
$$y = 6$$

ANSWER: $y = 6$

Check by substituting 6 for y.

Check: $2(6) - 6 = (-3)(6) + 24$
$12 - 6 = -18 + 24$
$\checkmark 6 = 6$

Solve each equation.

1. $4x = 3x + 6$ $5y = 2y + 9$ $7z = 5z + 18$
$4x - 3x = 6$
$x = 6$

2. $3z = 12 - z$

$3z + z = 12$

$4z = 12$

$z = \dfrac{12}{4}$

$z = 3$

$5a = 24 - 3a$

$7x = -27 - 2x$

3. $5x - 4 = 3x + 12$

$5x - 3x - 4 = 12$

$2x - 4 = 12$

$2x = 12 + 4$

$2x = 16$

$x = \dfrac{16}{2}$

$x = 8$

$9b - 7 = 5b + 25$

$8y - 9 = -y + 9$

4. $4m = 3m + 8$

$8x = 6x + 5$

$9y = 6y - 12$

5. $12x = 7x + 20$

$9y = 33 - 2y$

$3z = z - 18$

6. $4a + 17 = a - 13$

$6b - 12 = 2b - 4$

$z + 1 = \dfrac{1}{2}z + 4$

7. $5y + 9 = -2y + 30$

$14 + 2x = x + 9$

$11n - 3 = 9n + 3$

Applying Your Skills: Multistep Equations

Circle the correct equation. Then find the value of the variable.

1. Eight times a number n plus 9 is equal to 73. Which equation can you use to find n?

   a. $8n - 9 = 73$
   b. $8n + 9 = 73$
   c. $9n + 8 = 73$
   d. $8(n + 9) = 73$

   $n =$

2. Five times a number w minus 7 is equal to three times the same number plus 19. Which equation is true for w?

   a. $5w + 7 = 3w + 19$
   b. $5w - 7 = 3w - 19$
   c. $5w - 7 = 3w + 19$
   d. $5(w + 7) = 3(w + 19)$

   $w =$

3. Susan and Terry run a day-care center at Susan's house. Susan earns twice as much as Terry. If they earn $2,700 combined during May, which equation can be used to find Terry's share t?

   a. $2t - t = \$2,700$
   b. $t - 2t = \$2,700$
   c. $3t - t = \$2,700$
   d. $t + 2t = \$2,700$

   $t =$

4. As an appliance sales person, you earn $1,850 each month in salary. You also receive an additional $25 for each appliance you sell. If last month you earned a total of $2,375, how many appliances a did you sell?

   a. $\$1,850 - \$25a = \$2,350$
   b. $\$25a + \$2,375 = \$1,850$
   c. $\$25a + \$1,850 = \$2,375$
   d. $\$25a + \$1,850a = \$2,375$

   $a =$

Write an equation for each problem. Then solve your equation to find the variable.

5. A number n multiplied by 3 is equal to the same number added to 12. What is the number n?

   Equation:

   $n =$

6. Two-thirds of a number m plus one-sixth of the same number is equal to 25. What is the number m?

   Equation:

   $m =$

7. Suppose you pay a 6% state sales tax. You decide to buy a new bed for a total cost of $630.70, including tax. What is the price p of the bed before tax?

Equation:

$p =$

8. Joe has money in a savings account. If he adds $50 a month for 6 months, he will have three times the amount he has now, not counting the interest. How much money m does Joe have in his account now?

Equation:

$m =$

Write an equation for each drawing. Solve your equation. (Note: More than one equation is possible, but only one solution is correct.)

9.

Equation:

$x =$

10.

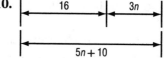

Equation:

$n =$

11.

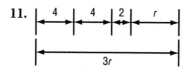

Equation:

$r =$

12.

Equation:

$y =$

Find the value of x in each triangle below. (Remember: $P = s_1 + s_2 + s_3$)

13.

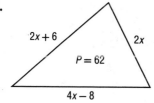

$x =$

14.

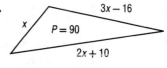

$x =$

Solving an Equation with Parentheses

Parentheses are commonly used in algebraic equations. Parentheses are used to identify terms (numbers or variables) that are to be multiplied by another term, often a number.

The first step in solving an equation is to remove the parentheses by multiplication. Then, combine separated variables and solve for the variable.

Follow these four steps to solve an algebraic equation.

STEP 1 Remove parentheses by distributing the multiplication.

STEP 2 Combine separated variables.

STEP 3 Do all addition or subtraction.

STEP 4 Do all multiplication or division.

To remove parentheses, multiply each term inside the parentheses by the term outside the parentheses—this is called *distributing the multiplication*. If the parentheses are preceded by a negative sign or number, remove the parentheses by distributing the −1 to each term within the parentheses.

(**Note:** Recall that when no number is written before parentheses, the number implied is 1. When parentheses are preceded by a negative sign, it is actually −1.)

EXAMPLES
$$4(x + 3) = 4x + 12$$
$$5(y - 4) = 5y - 20$$
$$-3(n + 3) = -3n - 9$$
$$-2(z - 1) = -2z + 2$$

EXAMPLE 1 Solve for x: $4(x + 3) = 20$　　　　**Solve:**　$4(x + 3) = 20$

STEP 1 Remove parentheses by multiplying each term.　　　　$4x + 12 = 20$

STEP 2 Subtract 12 from 20.　　　$4x = 20 - 12$
　　　　　　　　　　　　　　　　　　$4x = 8$

STEP 3 Divide 8 by 4.　　　　　　$x = \frac{8}{4}$
　　　　　　　　　　　　　　　　　　$\boldsymbol{x = 2}$

ANSWER: $x = 2$

EXAMPLE 2 Solve for z: $4z - (3z + 2) = 5$　　　**Solve:** $4z - (3z + 2) = 5$

STEP 1 Remove parentheses by distributing the −1 to each term within the parentheses.　　　$4z - 3z - 2 = 5$

STEP 2 Combine the z terms.　　　　　$z - 2 = 5$

STEP 3 Add 2 to 5.　　　　　　　　　$z = 5 + 2$
　　　　　　　　　　　　　　　　　　　$\boldsymbol{z = 7}$

ANSWER: $z = 7$

Solve each equation. The first problem in each row is done as an example.
(Remember: Multiply each term inside the parentheses by the coefficient
to remove the parentheses as your first step.)

1. $2(x + 4) = 20$
$$2x + 8 = 20$$
$$2x = 20 - 8$$
$$2x = 12$$
$$x = \frac{12}{2}$$
$$x = 6$$

$3(n + 2) = 27$

$5(b + 3) = 40$

2. $3(y - 6) = 15$
$$3y - 18 = 15$$
$$3y = 15 + 18$$
$$y = \frac{33}{3}$$
$$y = 11$$

$2(a - 3) = 16$

$4(b - 2) = 8$

3. $-2(x - 4) = -12$
$$-2x + 8 = -12$$
$$-2x = -12 - 8$$
$$-2x = -20$$
$$x = \frac{-20}{-2}$$
$$x = 10$$

$-3(m - 2) = -15$

$-2(2y + 1) = 6$

4. $4(y - 3) = 3(y + 6)$
$$4y - 12 = 3y + 18$$
$$4y - 3y - 12 = 18$$
$$y - 12 = 18$$
$$y = 18 + 12$$
$$y = 30$$

$5(z - 1) = 4(z + 4)$

$7(a - 1) = 6(a + 1)$

Applying Your Skills: Equations with Parentheses

In the following examples, notice how parentheses are used to represent information contained in word problems. Study these examples before solving the word problems on the next page.

EXAMPLE 1 Three times the quantity of a number minus 4 is equal to two times the sum of the number plus 3. What is the number?

STEP 1 Let x = the unknown number

$3(x - 4)$ is three times the quantity of x minus 4

$2(x + 3)$ is two times the sum of x plus 3

STEP 2 Write an equation for the problem. **Solve: $3(x - 4) = 2(x + 3)$**

STEP 3 Solve the equation.

a. Multiply to remove parentheses. $\qquad$ $3x - 12 = 2x + 6$

b. Move $2x$ to the left side of the equation, changing the sign from + to –. $\qquad$ $3x - 2x - 12 = 6$

c. Combine like terms. $\qquad$ $x - 12 = 6$

d. Move 12 to the right side of the equation. $\qquad$ $x = 6 + 12$

e. Add 12 to 6. $\qquad$ **$x = 18$**

ANSWER: $x = 18$

EXAMPLE 2 Lin, Anne, and Sarah share living expenses. Anne pays $25 less rent than Lin. Sarah pays twice as much rent as Anne. If the total rent is $1,425, how much rent does each person pay?

STEP 1 Let x = Lin's rent

$x - 25$ = Anne's rent

$2(x - 25)$ = Sarah's rent

(**Hint:** Since you don't know how much Lin pays for rent, let her rent equal x.)

STEP 2 Write an equation. $\qquad$ Lin's + Anne's + Sarah's = total rent

Solve: $x + (x - 25) + 2(x - 25) = 1,425$

STEP 3 Solve the equation.

a. Multiply to remove parentheses. $\qquad$ $x + x - 25 + 2x - 50 = 1,425$

b. Combine like terms. $\qquad$ $4x - 75 = 1,425$

c. Move 75 to the right side of the equation. $\qquad$ $4x = 1,425 + 75$

d. Add 75 to 1,425. $\qquad$ $4x = 1,500$

e. Divide 1,500 by 4. $\qquad$ $x = \dfrac{1,500}{4}$

$\qquad$ **$x = 375$**

ANSWER: $\quad$ $x = \$375$, **Lin's rent**

$x - 25 = \$350$, **Anne's rent**

$2(x - 25) = \$700$, **Sarah's rent**

Circle the correct equation. Then find the value of the variable.

1. Three times the sum of a number n plus 1 equals two times the sum of the number plus 4. Which equation can be used to find n?

   a. $3(n + 1) = 2n + 4$
   b. $3(n + 1) = 2(n + 4)$
   c. $3(n + 4) = 2(n + 1)$
   d. $3n + 1 = 2n + 4$

   $n =$

2. As a restaurant manager you employ three cooks. They earn a total of $4,400 each month. Earl earns $60 less than James. Maria earns twice as much as Earl. Which equation can be used to find James' monthly earnings j?

   a. $j - (j + \$60) + 2(j + \$60) = \$4,400$
   b. $j + (j - \$60) - 2(j + \$60) = \$4,400$
   c. $j + (j - \$60) + 2(j - \$60) = \$4,400$
   d. $j + j + \$60 + 2j - \$60 = \$4,400$

   $j =$

Write an equation for each problem. Then solve your equation to find the variable.

3. Four times the sum of a number n plus 2 equals 3 times the sum of the number plus 5. What is n?

   Equation:

   $n =$

4. Two-thirds the quantity y minus 3 is equal to one-third y. What is y?

   Equation:

   $y =$

5. As a school aide you bought chalk, erasers, and rulers for the class. You paid $8.50 more for the rulers than for the erasers. You paid $4.75 more for the erasers than for the chalk. If the bill is $72, what did each item cost? (**Hint:** Write an equation for the cost c of chalk.)

   Equation:

   Chalk:
   Erasers:
   Rulers:

6. Consecutive integers are whole numbers that follow one another. For example, 11, 12, and 13 are consecutive integers. If the sum of three consecutive integers is 54, what are the integers? (**Hint:** Let $x = $ 1st integer)

   Equation:

   1st integer:
   2nd integer:
   3rd integer:

Rearranging a Formula

The skills you use in solving equations can be used to rearrange a formula.
To *rearrange a formula* is to solve for a different variable.

EXAMPLE 1 Formula for the circumference C of a circle: $C = 2\pi r$

When solving a circumference problem, you are usually told the radius r.

Knowing r, you are asked to find C.

Suppose, though, you are told C. You are then asked to find r.

To Do   Rearrange $C = 2\pi r$ to solve for r.

STEP 1   Divide each side of the formula by 2π. $\dfrac{C}{2\pi} = \left(\dfrac{2\pi}{2\pi}\right)r$

STEP 2   On the right side, discard $\dfrac{2\pi}{2\pi}$ ($= 1$). $\dfrac{C}{2\pi} = r$

ANSWER: $r = \dfrac{C}{2\pi}$ The radius r equals C divided by 2π.

EXAMPLE 2 Formula for the perimeter P of a rectangle: $P = 2(l + w)$

Suppose you know P and the length l and want to find the width w.

To Do   Rearrange $P = 2(l + w)$ to solve for w.

STEP 1   Divide each side of the formula by 2. $\dfrac{P}{2} = \left(\dfrac{2}{2}\right)(l + w)$

Discard $\dfrac{2}{2}$ ($= 1$). Also, discard the parentheses; now they are not needed. $\dfrac{P}{2} = l + w$

STEP 2   Subtract l from each side. $\dfrac{P}{2} - l = w$

ANSWER: $w = \dfrac{P}{2} - l$ The width w equals *half the perimeter* minus the length l.

EXAMPLE 3 Formula for the area A of a circle: $A = \pi r^2$

Suppose you know A and want to find r.

To Do   Rearrange $A = \pi r^2$ to solve for r.

STEP 1   Divide each side of the formula by π. $\dfrac{A}{\pi} = \left(\dfrac{\pi}{\pi}\right)r^2$

Discard $\dfrac{\pi}{\pi}$ ($= 1$). $\dfrac{A}{\pi} = r^2$

STEP 2   Take the square root of each side. $\sqrt{\dfrac{A}{\pi}} = \sqrt{r^2} = r$

ANSWER: $r = \sqrt{\left(\dfrac{A}{\pi}\right)}$ The radius r equals the square root of *the area A divided by* π.

Applying Your Skills: Rearranging a Formula

1. The formula for the circumference of a circle is $C = 2\pi r$.

   a. Rearrange this formula to show how to find r when you know C.

   b. A circular wading pool has a circumference of 98 feet. What is the approximate radius of this pool? (Use $\pi \approx 3.14$.)

2. Muriel surveyed a triangular plot of land. The number she wrote for one of the lengths got smeared and now she can't read it. She does know that the perimeter is 215 yards, the shortest side is 57 yards, and the longest side is 88 yards.

   a. The formula for the perimeter of a triangle is $P = a + b + c$, where a, b, and c are the lengths of the sides of the triangle. Rearrange this formula to show how to find c when you know P, a, and b.

   b. Using the formula found in part a, find the length of the third side.

3. A rectangular storage room has a volume of 1,584 cubic feet.

   a. Rearrange the volume formula $V = lwh$ to solve for h.

   b. Find h when $l = 22$ feet and $w = 9$ feet.

4. Trevor is installing a triangular window in a new restaurant.

   a. The formula for the area of a triangle is $A = \frac{1}{2}bh$. Rearrange this formula to find the height h when you know area A and base b.

   b. Using the formula found in part a, find h when $A = 12$ square feet and $b = 6$ feet.

5. Colby, a farmer, filled his silo with grain. The silo is in the shape of a cylinder and has a radius r of 7 feet.

   a. The formula for the volume of a cylinder is $V = \pi r^2 h$. Rearrange this formula to show how to find h when you know V and r.

   b. If the silo has a volume of about 4,000 cubic feet, what is the approximate height h of the silo? (Use $\pi \approx \frac{22}{7}$.)

6. Simple interest i paid on a loan is given by the formula $i = prt$, where

   p = principal (amount borrowed)
   r = loan rate (yearly)
   t = time (number of years)

   Rearrange $i = prt$ to show how to find r when you know i, p, and t.

Multistep Equations Review

Solve the problems below. When you finish, check your answers at the back of the book. Then correct any errors.

Basic Equations

1. $3x + 14 = 62$

2. $4n + 9 = 25$

3. $r - 7 = 65$

4. $4y - 13 = 31$

5. $\frac{x}{2} + 5 = 17$

6. $\frac{2}{3}n - 4 = 12$

Equations with Separated Variables

7. $4n + 3n = 35$

8. $5x - 2x + 9 = 36$

9. $\frac{2}{3}y - \frac{1}{6}y + 4 = 19$

Equations with Terms on Both Sides

10. $7x = 3x + 20$

11. $4n + 7 = 3n + 30$

12. $11y - 6 = 9y + 28$

Equations with Parentheses

13. $6(b - 2) = 12$

14. $4(z + 2) = 2z + 40$

15. $3(y + 2) = 2(y + 5)$

Mixed Practice

16. $5w - 2w = 39$

17. $3x + 5 = 20$

18. $4n = 2n + 14$

19. $5(x - 3) = 10$

20. $5s - 2s + 9 = 24$

21. $6r + 2 = 3r + 11$

22. $2(x + 5) = x + 4$

23. $4y - 8 = 32$

24. $\frac{n}{2} = n + 1$

25. $4(y + 3) = 2y$

26. $3(z - 2) = 2(z - 1)$

27. $\frac{3}{4}x - \frac{1}{4}x + 5 = 27$

28. $4p = 2p + 8$

29. $5n - 6 = 2n + 12$

30. $x - 2 = \frac{1}{2}x + 5$

Circle the correct equation. Then find the value of the variable.

31. Six times a number m subtract 4 is equal to 26. Which equation can you use to find m?

    a. $6m - 26 = 4$
    b. $4m - 26 = 4$
    c. $4m - 6 = 26$
    d. $6m - 4 = 26$

    $m =$

32. If Jolene adds five stamps each month for 8 months to her collection, she will have twice the number of stamps she has now. Which equation can you use to find the number n of stamps Jolene has now?

    a. $n - 40 = 2n$
    b. $2n + 8(5) = n$
    c. $n + 8(5) = 2n$
    d. $5 + 8 = 2n$

    $n =$

Write an equation for each problem. Then solve your equation to find the variable.

33. Five times the quantity x minus 1 equals 3 times the quantity x plus 9. What is x?

Equation:

    $x =$

34. Four times a number n minus 3 times the sum of the number plus 2 is equal to 5. What is the number?

Equation:

    $n =$

35. Lateisha has a coin collection. If she adds five coins each month, the size of her collection will triple in 2 years. How many coins c does Lateisha have in her collection now?

Equation:

Number of coins:

36. Alyce, Jon, and Kim worked together on a painting job. Jon worked 8 fewer hours than Alyce. Kim worked twice as many hours as Jon. If the total hours worked on the job was 72, how many hours did each person work? (**Hint:** Let $h =$ hours worked by Alyce)

Equation:

Alyce:
Jon:
Kim:

37. Consecutive even integers are even whole numbers that follow one another. For example, 12, 14, and 16 are consecutive even integers. If the sum of three consecutive even integers is 66, what are the integers? (**Hint:** Let x = 1st number)

Equation:

1st number:

2nd number:

3rd number:

38. Find the value of x in the triangle below.

Equation:

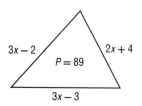

$x =$

Write an equation for each drawing. Solve your equation. (Note: More than one equation is possible, but only one solution is correct.)

39.

Equation:

$x =$

40.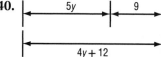

Equation:

$y =$

41.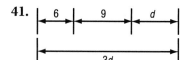

Equation:

$d =$

42.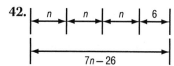

Equation:

$n =$

SPECIAL EQUATIONS

Understanding Ratios

A **ratio** is a comparison of two numbers. For example, if there are 5 women and 4 men in a class, the *ratio of women to men* is 5 to 4.

You can write the ratio *5 to 4*:

- as a fraction, *5 to 4* is $\frac{5}{4}$.
- using a colon, *5 to 4* is 5:4.

Read the ratios $\frac{5}{4}$ and 5:4 as "five to four."

Write a ratio in the same order as it appears in a question. In the example above, the ratio of men to women is 4 to 5.

To simplify a ratio, write it as a fraction and follow these three rules.

1. Simplify a ratio to lowest terms.
$$5 \text{ to } 10 = \frac{5}{10} = \frac{1}{2}$$

2. If the ratio is an improper fraction, leave it as an improper fraction.
$$8 \text{ to } 6 = \frac{8}{6} = \frac{4}{3}$$

3. Write a whole-number ratio as an improper fraction.
$$4 \text{ to } 1 = \frac{4}{1}$$

Write a fractional ratio for each problem. Simplify when possible.

1. There are 12 girls and 16 boys in Kira's class. What is the ratio of boys to girls?

2. Chad's grandfather is 80 years old. Chad is 25 years old. What is the ratio of Chad's age to his grandfather's age?

3. Tiffani's soccer team won 12 games and lost 8.

   a. What is the ratio of games won to games lost?
   b. What is the ratio of games won to games played?

4. Of the 18 teachers in Shane's school, 14 are women. What is the ratio of male teachers to female teachers at this school?

5. What is the ratio of the distance between Rose City and Chad to the distance between Rose City and Garrett?

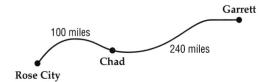

Applying Your Skills: Using Equations to Solve Ratio Problems

Some ratio problems are most easily solved using equations.

EXAMPLE The ratio of two numbers is 4 to 3. The sum of the numbers is 42. What are the two numbers?

STEP 1 Represent the greater number as $4x$ and the lesser number as $3x$.

STEP 2 Set the sum of the numbers equal to 42.

$$4x + 3x = 42$$

STEP 3 Solve the equation for x.

$$4x + 3x = 42$$
$$7x = 42$$
$$x = \frac{42}{7}$$
$$x = 6$$

STEP 4 Determine the two numbers.

greater number: $4x = 4(6) =$ **24**

lesser number: $3x = 3(6) =$ **18**

> The key to solving this problem is choosing to represent the unknown numbers as $4x$ and $3x$—whole number terms that are in the ratio of 4 to 3.

ANSWER: 18 and 24

Write an equation for each problem. Then solve your equation to find the unknown numbers.

1. The ratio of two numbers is 3 to 2. The sum of the numbers is 75. What are the two numbers?

   Equation:

   first number:
   second number:

2. The ratio of two numbers is 5 to 2. The difference of the numbers is 18. What are the two numbers?

   Equation:

   first number:
   second number:

3. Suppose you and Tomas start a house-painting business. For every $10 you earn, Tomas earns $12. On a job that pays $2,420, how much will each of you earn?

   Equation:

   You:
   Tomas:

4. Your boss at the nursery asks you to design a rectangular garden. The ratio of the length to the width is to be 7 to 4; the perimeter is to be 176 feet. What length and width should you choose?

   Equation:

   length:
   width:

Understanding Proportions

A **proportion** is an equation made up of two equal ratios. For example, if you add 2 cups of orange juice to 3 cups of pineapple juice, the ratio of orange juice to pineapple juice is $\frac{2}{3}$. You make the same drink by adding 4 cups of orange juice to 6 cups of pineapple juice. The drinks are the same because the ratios of orange juice to pineapple juice are equal: $\frac{2}{3} = \frac{4}{6}$.

You read a proportion as two equal ratios connected by the word *as*.

$\frac{2}{3} = \frac{4}{6}$ is read "2 is to 3 *as* 4 is to 6."

> A proportion is usually written as a pair of equivalent (equal) fractions.

In a proportion, the cross products are equal. To find the cross products, multiply each numerator by the opposite denominator.

Cross Multiplication

$$\frac{2}{3} \underset{\longleftarrow}{\overset{\longrightarrow}{\times}} \frac{4}{6}$$

Equal Cross Products

$$2(6) = 3(4)$$
$$12 = 12$$

Finding a Missing Number in a Proportion

To complete a proportion, select a variable for the missing number.

- Write the cross products.

- Solve the equation for the chosen variable.

EXAMPLE 1 Find the missing number: $\frac{n}{3} = \frac{8}{12}$

STEP 1 Cross multiply. $12n = 24$

STEP 2 Divide each side by 12. $n = \frac{24}{12} = \mathbf{2}$

ANSWER: $n = \mathbf{2}$

EXAMPLE 2 Find the missing number: $\frac{4}{x} = \frac{20}{15}$

STEP 1 Cross multiply. $20x = 60$

STEP 2 Divide each side by 20. $x = \frac{60}{20} = \mathbf{3}$

ANSWER: $x = \mathbf{3}$

Cross multiply to see if each pair of fractions forms a proportion.
Circle Yes if the cross products are equal, No if they are not.

1. $\dfrac{2}{3} \overset{?}{=} \dfrac{9}{12}$ Yes No
$\qquad\qquad\qquad$ $\dfrac{3}{4} \overset{?}{=} \dfrac{6}{8}$ Yes No
$\qquad\qquad\qquad$ $\dfrac{4}{5} \overset{?}{=} \dfrac{5}{4}$ Yes No

2. $\dfrac{3}{7} \overset{?}{=} \dfrac{9}{21}$ Yes No
$\qquad\qquad\qquad$ $\dfrac{9}{16} \overset{?}{=} \dfrac{5}{8}$ Yes No
$\qquad\qquad\qquad$ $\dfrac{4}{5} \overset{?}{=} \dfrac{8}{10}$ Yes No

Write cross products for each proportion. The first problem is done as an example.

3. $\dfrac{n}{2} = \dfrac{8}{16}$
$\qquad$ $16n = 2 \times 8$
$\qquad$ $16n = 16$
$\qquad\qquad$ $\dfrac{4}{12} = \dfrac{3}{n}$
$\qquad\qquad\qquad$ $\dfrac{8}{20} = \dfrac{x}{5}$
$\qquad\qquad\qquad$ $\dfrac{x}{3} = \dfrac{10}{12}$

4. $\dfrac{10}{x} = \dfrac{5}{9}$
$\qquad\qquad$ $\dfrac{n}{4} = \dfrac{24}{32}$
$\qquad\qquad\qquad$ $\dfrac{8}{3} = \dfrac{x}{6}$
$\qquad\qquad\qquad$ $\dfrac{n}{8} = \dfrac{18}{12}$

In each proportion below, find the value of n or x.

5. $\dfrac{n}{2} = \dfrac{7}{14}$
$\qquad\qquad$ $\dfrac{x}{3} = \dfrac{12}{18}$
$\qquad\qquad\qquad$ $\dfrac{6}{8} = \dfrac{x}{4}$
$\qquad\qquad\qquad$ $\dfrac{5}{n} = \dfrac{10}{4}$

$\quad$ $n =$
$\qquad\qquad\qquad$ $x =$
$\qquad\qquad\qquad$ $x =$
$\qquad\qquad\qquad$ $n =$

6. $\dfrac{n}{6} = \dfrac{27}{18}$
$\qquad\qquad$ $\dfrac{3}{8} = \dfrac{x}{32}$
$\qquad\qquad\qquad$ $\dfrac{n}{10} = \dfrac{4}{5}$
$\qquad\qquad\qquad$ $\dfrac{4}{x} = \dfrac{12}{9}$

$\quad$ $n =$
$\qquad\qquad\qquad$ $x =$
$\qquad\qquad\qquad$ $n =$
$\qquad\qquad\qquad$ $x =$

Applying Your Skills: Using Proportions

Proportions are used to solve word problems involving comparisons or rates. When setting up a proportion, remember to write the terms of the ratios in the order stated in the problem. As you will see, numbers in a proportion are not always whole numbers.

EXAMPLE 1 A recipe calls for 2 cups of sugar for each 5 cups of flour. How many cups of sugar are needed for 14 cups of flour?

STEP 1 Write a proportion where each ratio is

$$\frac{\text{cups of sugar}}{\text{cups of flour}}.$$

Let n stand for the unknown cups of sugar.

$$\frac{2}{5} = \frac{n}{14}$$

STEP 2 Write the cross products.

$$5n = 28$$

STEP 3 Divide by 5 to solve for n.

$$n = \frac{28}{5} = 5\frac{3}{5}$$

ANSWER: $n = 5\frac{3}{5}$ **cups of sugar**

EXAMPLE 2 Working 7 hours, Blake earned $64. At this same pay rate, how much would Blake earn working 10 hours?

STEP 1 Write a proportion where each ratio is

$$\frac{\text{hours worked}}{\text{amount earned}}.$$

Let n stand for the unknown earnings.

$$\frac{7}{\$64} = \frac{10}{n}$$

STEP 2 Write the cross products.

$$7n = \$640$$

STEP 3 Divide by 7 to solve for n.

$$n = \frac{\$640}{7} = \mathbf{\$91.43}$$
(to the nearest cent)

ANSWER: $n = \mathbf{\$91.43}$

Circle the proportion that can be used to solve each problem.

1. On a map of Ohio, an actual distance of 75 miles is only 2 inches of map distance. On this map, how many inches n apart are two towns whose actual distance apart is 280 miles?

   a. $\dfrac{n}{2} = \dfrac{75}{280}$

   b. $\dfrac{75}{2} = \dfrac{280}{n}$

   c. $\dfrac{n}{75} = \dfrac{2}{280}$

   d. $\dfrac{75}{2} = \dfrac{n}{280}$

2. When he painted his house, Bob mixed 2 gallons of blue paint with 3 gallons of white. How many gallons g of blue paint did Bob use if he used a total of 19 gallons of white paint?

   a. $\dfrac{g}{3} = \dfrac{2}{19}$

   b. $\dfrac{g}{2} = \dfrac{3}{19}$

   c. $\dfrac{2}{3} = \dfrac{g}{19}$

   d. $\dfrac{2}{g} = \dfrac{3}{19}$

3. Bonita drove 450 miles in 8 hours. At this same rate, how many hours t would it take Bonita to drive another 225 miles?

   a. $\dfrac{450}{8} = \dfrac{225}{t}$

   b. $\dfrac{t}{8} = \dfrac{450}{225}$

   c. $\dfrac{t}{450} = \dfrac{8}{225}$

   d. $\dfrac{450}{8} = \dfrac{t}{225}$

4. In a survey of 280 high school students, 3 out of 5 said they plan to work during summer vacation. How many students n do *not* plan to work?

   a. $\dfrac{3}{5} = \dfrac{n}{280}$

   b. $\dfrac{n}{280} = \dfrac{3}{8}$

   c. $\dfrac{2}{5} = \dfrac{n}{280}$

   d. $\dfrac{2}{3} = \dfrac{n}{280}$

Identify the variable, write a proportion and solve.
(Note: More than one correct proportion is possible, but only one answer is correct.)

5. Suppose you work in a deli. If you charge $2.24 for 8 ounces of sliced beef, how much should you charge for 12 ounces?

6. Jeni drove her car 102 miles on 3 gallons of gas. How far can Jeni expect to drive on 16 gallons of gas?

7. A light tan paint is made by mixing 2 parts of brown paint to 5 parts of white. How many parts of brown must be mixed with 15 parts of white to make the light tan color?

8. If 12.7 centimeters exactly equal 5 inches, how many centimeters exactly equal 2 inches?

9. You are offered a job that pays $76.80 for 8 hours of work. At this rate how much can you earn if you work a total of 28 hours?

10. Epoxy is to be mixed in a ratio of 3 parts hardener to 7 parts base. How many drops of hardener are needed for 28 drops of base.

Solving a Linear Equation

A **linear equation** is an equation that contains two variables where each variable has an exponent of 1. (**Remember:** An exponent of 1 is usually not written.) For example, $y = 2x + 3$ is a linear equation that contains the variables x and y.

A linear equation has more than one solution. Each solution has two parts: an x-value and a y-value. For each value of x, there is one matching value of y. In $y = 2x - 3$, if you substitute 5 for x, y is equal to 7 ($2(5) - 3$).

To solve a linear equation follow these two steps.

STEP 1   Choose several values for the second variable.

STEP 2   Substitute each value for the second variable into the equation and find the matching value of the first variable.

To keep the solutions in neat order, you write them in a **Table of Values.**

EXAMPLE 1   Solve for y: $y = 3x$, when $x = -2, 0, 2,$ and 4

STEP 1   Write the chosen values for x in the Table of Values.

STEP 2   Substitute each value for x into the equation $y = 3x$.

STEP 3   Fill in the Table of Values for y.

x value	$y = 3x$
$x = -2$	$y = 3(-2) = \mathbf{-6}$
$x = 0$	$y = 3(0) = \mathbf{0}$
$x = 2$	$y = 3(2) = \mathbf{6}$
$x = 4$	$y = 3(4) = \mathbf{12}$

Table of Values	
x	**y**
-2	-6
0	0
2	6
4	12

EXAMPLE 2   Solve for y: $y = 2x - 3$, when $x = -1, 0,$ and 3

STEP 1   Write the chosen values of x in the Table of Values.

STEP 2   Substitute each value of x into the equation $y = 2x - 3$.

STEP 3   Fill in the Table of Values for y.

x value	$y = 2x - 3$
$x = -1$	$y = \overset{-2}{\overbrace{2(-1)}} - 3 = \mathbf{-5}$
$x = 0$	$y = \overset{0}{\overbrace{2(0)}} - 3 = \mathbf{-3}$
$x = 3$	$y = \overset{6}{\overbrace{2(3)}} - 3 = \mathbf{3}$

Table of Values	
x	**y**
-1	-5
0	-3
3	3

For each linear equation, complete the Table of Values.

1. $y = x + 4$

Table of Values	
x	**y**
0	
1	
2	
3	

2. $y = 3x - 2$

Table of Values	
x	**y**
1	
2	
3	
4	

3. $a = -2b + 3$

Table of Values	
b	**a**
−1	
0	
1	
2	

4. $r = \frac{1}{2}s + 3$

Table of Values	
s	**r**
−2	
0	
2	
4	

Complete the Table of Values as indicated.

5. The perimeter of the rectangle below is given by the equation $P = 2w + 6$. Complete the Table of Values for P.

Table of Values	
w	**P**
6	
10	
15	
20	

6. M and N are complementary angles. Use the equation $m\angle M = 90° - m\angle N$ to complete the Table of Values for $m\angle M$.

Table of Values	
m∠N	**m∠M**
25°	
45°	
62°	
78°	

Solving a System of Equations

Two linear equations often have a **common solution**—an *x*- and a *y*-value that are a solution for each equation at the same time. Finding a common solution for two or more linear equations is called **solving a system of equations.** One way to find a common solution is to subtract one equation from the other.

EXAMPLE 1 What *x*-value and *y*-value are a common solution to the following system of equations?

Equation 1: $y = 2x + 3$
Equation 2: $y = x + 2$

STEP 1 Subtract Equation 2 from Equation 1. Subtract the *y* terms on the left side. Subtract the *x* terms and numbers on the right side.

$$\begin{aligned} y = 2x + 3 &\rightarrow y = 2x + 3 \\ -(y = \ x + 2) &\rightarrow \underline{-y = -x - 2} \\ & \qquad 0 = \ x + 1 \end{aligned}$$

STEP 2 Solve for *x* by subtracting 1 from each side of the equation.

$$0 - 1 = x + 1 - 1$$
$$-1 = x$$
or $x = -1$

STEP 3 Substitute the *x*-value (−1) into the second equation. Solve the equation to find the *y*-value.

$$y = x + 2$$
$$= -1 + 2$$
$$y = 1$$

ANSWER: *x* = −1, *y* = 1

Check:

Equation 1	Equation 2
$y = 2x + 3$	$y = x + 2$
$1 = 2(-1) + 3$	$1 = -1 + 2$
$1 = -2 + 3$	✓$1 = 1$
✓$1 = 1$	

EXAMPLE 2 What *x*-value and *y*-value are a common solution to the following equations?

Equation 1: $y = 2x + 1$
Equation 2: $y = -x + 4$

STEP 1 Subtract Equation 2 from Equation 1.

$$\begin{aligned} y = 2x + 1 &\rightarrow y = \ 2x + 1 \\ -(y = -x + 4) &\rightarrow \underline{-y = -(-x) - 4} \\ & \qquad 0 = \ 3x - 3 \end{aligned}$$

STEP 2 Solve for *x* by adding 3 to each side of the equation. Then divide each side by 3.

$$0 + 3 = 3x - 3 + 3$$
$$3 = 3x \text{ (or } 3x = 3)$$
$$\frac{3x}{3} = \frac{3}{3} \text{ (Divide by 3.)}$$
$$x = 1$$

STEP 3 Substitute the *x*-value (1) into the second equation. Solve the equation to find the *y*-value.

$$y = -x + 4$$
$$= -1 + 4$$
$$y = 3$$

ANSWER: *x* = 1, *y* = 3

Check:

Equation 1	Equation 2
$y = 2x + 1$	$y = -x + 4$
$3 = 2(1) + 1$	$3 = -1 + 4$
$3 = 2 + 1$	✓$3 = 3$
✓$3 = 3$	

Use subtraction to solve each system of equations.

1. $y = 2x + 3$
   $y = x + 1$

   $x =$
   $y =$

2. $y = 3x + 6$
   $y = 2x + 1$

   $x =$
   $y =$

3. $y = 5x - 3$
   $y = 2x$

   $x =$
   $y =$

4. $y = 4x - 1$
   $y = 3x + 1$

   $x =$
   $y =$

5. $y = 2x + 9$
   $y = -x$

   $x =$
   $y =$

6. $y = x - 3$
   $y = -x + 5$

   $x =$
   $y =$

Quadratic Equations

In a linear equation, such as $y = 2x + 2$, there are many possible solutions—in fact, an infinite number.

- For each value of x there is a single value of y.
- For each value of y there is a single value of x.

A **quadratic equation,** unlike a linear equation, contains the square of a variable. For example, $y = x^2$ is a quadratic equation. For $y = x^2$, there are also an infinite number of solutions. When solving a quadratic equation, you sometimes want to find x-values when the y-value is given. As a first step, the variable that is squared is written to the left of the equals sign. For example, the equation $y = x^2$ would be written as $x^2 = y$.

Notice that for each possible value of y, there are *two correct* values of x. One value is a positive square root; the other is a negative square root.

EXAMPLE 1 Solve the equation $x^2 = y$ when $y = 16$.

STEP 1 Substitute 16 for the variable y.

$x^2 = 16$

STEP 2 Solve for x by taking the square root of each side of the equation.

$\sqrt{x^2} = \sqrt{16}$

$x = +4$ or -4

ANSWER: $x = \pm 4$

EXAMPLE 2 Solve the equation $3x^2 = y$ when $y = 12$.

STEP 1 Substitute 12 for the variable y.

$3x^2 = 12$

STEP 2 Divide each side of the equation by 3.

$\frac{3x^2}{3} = \frac{12}{3}$

$x^2 = 4$

STEP 3 Solve for x by taking the square root of each side.

$\sqrt{x^2} = \sqrt{4}$

$x = +2$ or -2

ANSWER: $x = \pm 2$

EXAMPLE 3 Solve the equation $2x^2 - 50 = y$ when $y = 0$.

STEP 1 Substitute 0 for the variable y.

$2x^2 - 50 = 0$

STEP 2 Add 50 to each side of the equation.

$2x^2 - 50 + 50 = 0 + 50$

$2x^2 = 50$

STEP 3 Divide each side of the equation by 2.

$\frac{2x^2}{2} = \frac{50}{2}$

$x^2 = 25$

STEP 4 Solve for x by taking the square root of each side of the equation.

$\sqrt{x^2} = \sqrt{25}$

$x = +5$ or -5

ANSWER: $x = \pm 5$

Solving for y

For each value of x, find the single correct value of y. The first problem in each row is done as an example.

1. $y = x^2$ when $x = 3$

   $y = 3^2$

     $= 9$

   $y = 2x^2$ when $x = -2$

   $y = \frac{3}{4}x^2$ when $x = 4$

2. $y = x^2 + 3$ when $x = 2$

   $y = 2^2 + 3$

     $= 4 + 3$

     $= 7$

   $y = x^2 - 7$ when $x = -3$

   $y = 2x^2 + 5$ when $x = 4$

Solving for x

For each value of y, find the *two* correct values of x. The first problem in each row is done as an example.

3. $x^2 = y$ when $y = 25$

   $x^2 = 25$

   $x = \pm 5$

   $3x^2 = y$ when $y = 48$

   $2x^2 = y$ when $y = 72$

4. $x^2 - 49 = y$ when $y = 0$

   $x^2 - 49 = 0$

       $x^2 = 49$

        $x = \pm 7$

   $x^2 - 64 = y$ when $y = 0$

   $4x^2 - 144 = y$ when $y = 0$

5. $x^2 + 4 = y$ when $y = 13$

   $x^2 + 4 = 13$

       $x^2 = 9$

        $x = \pm 3$

   $2x^2 = y$ when $y = 50$

   $x^2 - 50 = y$ when $y = 119$

Applying Your Skills: Special Equations

Linear Equations

1. Write a linear equation that tells the perimeter P of the rectangle below.

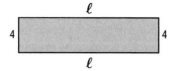

   $P =$

2. Suppose you start a business making and selling earrings. Your cost to make each pair of earrings is $1.25, and you sell the earrings for $9.95. Saturday you pay $120 to rent a sales booth at the county fair.

   Which equation shows the profit p you will make if you sells n pairs of earrings at the fair?

   a. $p = (\$9.95 + \$1.25)n - \$120$
   b. $p = (\$9.95 + \$1.25)n + \$120$
   c. $p = (\$9.95 - \$1.25)n - \$120$
   d. $p = (\$9.95 - \$1.25)n + \$120$

3. Sunshine Company produces solar heaters. The cost c of production is a linear function of the number n of heaters produced.

   $c = \$400n + \$5,000$

   a. What is the cost of producing 25 heaters?

   b. What is the *cost per heater* if 25 heaters are produced?

   c. What is the *cost per heater* if 50 heaters are produced?

Systems of Equations

4. At Frosty Yogurt the cost of a small frozen yogurt cone is $1 while the cost of a medium cone is $4. On Monday they sold 150 cones. A total of $465 was collected.

   Solve the equations below to find the number of each size sold.

   Let s = number of small cones
   m = number of medium cones

   1st equation: $s + 4m = 465$
   2nd equation: $s + m = 150$

5. A landscaper places two orders with a nursery. The first order is for 8 bushes and 10 small trees, at a total cost of $420. The second order is for 8 bushes and 4 small trees, at a total cost of $240.

Solve the equations below to find the price of a bush and the price of a tree.

Let b = price of a bush
t = price of a tree

1st equation: $8b + 10t = \$420$
2nd equation: $8b + 4t = \$240$

6. You are to buy a month's supply of dog food and cat food for an animal shelter. Dog food costs twice as much as cat food. The shelter has 40 cats and 20 dogs. Your food budget for this month is $680.

Solve the equations below to find how much you can spend for each type of food.

Let d = food cost per month per dog
c = food cost per month per cat

1st equation: $20d + 40c = \$680$
2nd equation: $d = 2c$
(Hint: First, rewrite the 1st equation as $20(2c) + 40c = 680$ and solve for c.)

Quadratic Equations

7. The formula for the surface area S_A of a sphere is a quadratic equation.

$S_A = 4\pi r^2$, where $\pi \approx 3.14$

Find the surface area of a spherical balloon with a radius of 4 inches. Round to the nearest whole number.

8. A water balloon is dropped from a bridge 96 feet above the water. An equation gives the height h in feet of the balloon above the water t seconds after the balloon is dropped.

$h = -16t^2 + 96$

a. After 2 seconds of falling ($t = 2$), what is the height of the balloon above the water?

b. How long does the balloon take to hit the water? ($h = 0$)

9. Ramey Manufacturing makes outdoor sheds. The profit p in dollars made from the Cabin model is approximated by an equation. Provided that the number n of sheds sold is less than 600, the equation is:

$p = 0.8n^2 + 50n - 6000$

What is the total profit during a month when 100 sheds are sold?

Special Equations Review

Solve the problems below. When you finish, check your answers at the back of the book. Then correct any errors.

1. The Jets baseball team won 25 games and lost 10.

    a. What is the ratio of games won to games lost?
    b. What is the ratio of games won to games played?

Write an equation for each problem. Then solve your equation to find the variable.

2. The ratio of two numbers is 3 to 4. The sum of the numbers is 63. What are the two numbers?

3. A soccer practice field is 3 times as long as it is wide. If the perimeter of the field is 176 yards, what is the length of the field?

Solve each proportion.

4. $\dfrac{n}{8} = \dfrac{9}{12}$

5. $\dfrac{3}{y} = \dfrac{18}{12}$

6. $\dfrac{25}{40} = \dfrac{20}{x}$

Write a proportion and solve each problem.

7. Regan earned $46.20 for 7 hours of work. At this rate, how much can Regan earn working 12 hours?

8. Three out of every 5 teachers at Jefferson High School are men. How many of Jefferson's 40 teachers are men?

For each linear equation, complete the Table of Values.

9. $y = x - 2$

Table of Values	
x	**y**
0	
2	
4	
6	

10. $y = 2x + 3$

Table of Values	
x	**y**
−1	
0	
1	
2	

Use subtraction to solve each system of equations.

11. $y = 2x + 5$
$\quad y = x + 3$

12. $y = 3x + 1$
$\quad y = x + 1$

$x =$
$y =$

$x =$
$y =$

For each value of x, find the single correct value of y.

13. $y = x^2$ when $x = 4$

14. $y = 2x^2$ when $x = -3$

For each value of y, find the *two* correct values of x.

15. $x^2 = y$ when $y = 36$

16. $4x^2 = y$ when $y = 64$

GRAPHING EQUATIONS

Becoming Familiar with a Coordinate Grid

A coordinate grid is formed by combining two number lines, one vertical and one horizontal.

- The vertical number line is called the **y-axis.**
- The horizontal number line is called the **x-axis.**
- The point at which the two axes meet is (0, 0) and is called the **origin.**

Every point on a coordinate grid has two coordinates, numbers that tell its position.

- The **x-coordinate** tells how far the point is from the y-axis.
  Positive x indicates the point is to the right of the y-axis.
  Negative x indicates the point is to the left of the y-axis.
- The **y-coordinate** tells how far the point is from the x-axis.
  Positive y indicates the point is above the x-axis.
  Negative y indicates the point is below the x-axis.
- Coordinates are usually written as an **ordered pair,** two numbers within parentheses. The x-coordinate (x value) is written first, followed by the y-coordinate (y value).

EXAMPLE   On the grid below, point A is (−6, 4).

x-coordinate ⌐ ⌐ y-coordinate

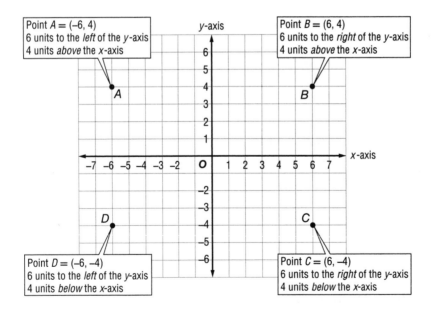

Point A = (−6, 4)
6 units to the *left* of the y-axis
4 units *above* the x-axis

Point B = (6, 4)
6 units to the *right* of the y-axis
4 units *above* the x-axis

Point D = (−6, −4)
6 units to the *left* of the y-axis
4 units *below* the x-axis

Point C = (6, −4)
6 units to the *right* of the y-axis
4 units *below* the x-axis

Write each pair of coordinates as an ordered pair. The first problem is done
as an example.

1. **a.** $x = 4$
   $y = -2$
   $(4, -2)$

   b. $x = -3$
   $y = 5$

   c. $x = 2$
   $y = 6$

   d. $x = -4$
   $y = -7$

Identify the x- and y-coordinates of each point. The first problem is done as
an example.

2. **a.** $R = (2, -8)$
   $x = 2$
   $y = -8$

   b. $S = (-1, 6)$
   $x =$
   $y =$

   c. $T = (0, 5)$
   $x =$
   $y =$

   d. $U = (-4, 0)$
   $x =$
   $y =$

Write the coordinates of each point as an ordered pair. The first point is done as
an example.

3.

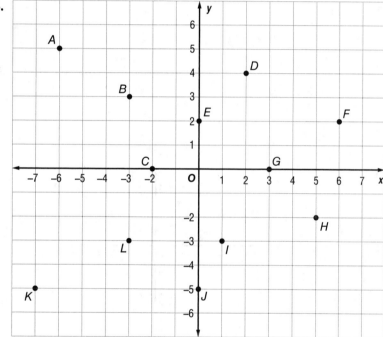

$A = (-6, 5)$ $B =$ $C =$ $D =$

$E =$ $F =$ $G =$ $H =$

$I =$ $J =$ $K =$ $L =$

Plotting Points on a Grid

To plot a point on a coordinate grid, follow these steps.

STEP 1 Locate the *x* value on the *x*-axis.

STEP 2 From the *x* value, move directly up (for a positive *y* value) or directly down (for a negative *y* value).

STEP 3 Label this point.

EXAMPLE 1 Plot point $A = (5, -4)$ on the grid at the right.

STEP 1 Find the value $x = 5$ on the *x*-axis.

STEP 2 From the value 5 on the *x*-axis, move down to the *y* value, −4.

STEP 3 Label point *A*.

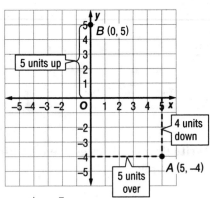

EXAMPLE 2 Plot point $B = (0, 5)$

STEP 1 Begin at $x = 0$, the *x* value. Move up to the *y* value, 5.

STEP 2 Label point *B*.

Plot and label each point.

1.

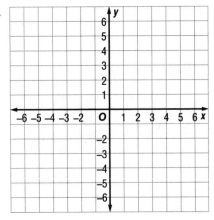

$A = (2, -3)$　　　　$B = (-4, 2)$

$C = (0, 4)$　　　　$D = (3, 5)$

2.

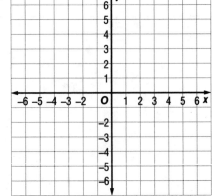

$E = (4, 6)$　　　　$F = (-1, -3)$

$G = (-5, 5)$　　　　$H = (2, 0)$

Graphing a Line

When plotted points lie on a straight line, draw a line through the points and extend this line to the edges of the grid. The points where the line crosses the axes are called **intercepts** of the line.

- The **x-intercept** is the point where a line crosses the x-axis. For the x-intercept, the y-coordinate is 0.
- The **y-intercept** is the point where a line crosses the y-axis. For the y-intercept, the x-coordinate is 0.

EXAMPLE Graph the line with an x-intercept of (−2, 0) and a y-intercept of (0, 2).

    STEP 1 Plot the points and label each point with its coordinates.

    STEP 2 Draw a line through the points. Extend the line to the edges of the grid.

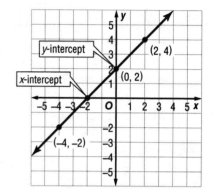

Graph the line containing each set of x- and y-intercepts.

1.

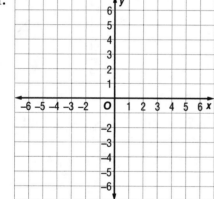

x-intercept = (2, 0)

y-intercept = (0, −6)

2.

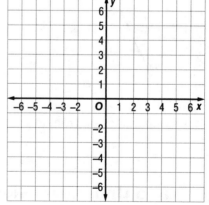

x-intercept = (0, 0)

y-intercept = (0, 0)

line also passes through (2, 2)

Finding the Slope of a Line

You are already familiar with the concept of **slope.** When you walk uphill, you walk up a slope. When you walk downhill, you walk down a slope. A graphed line, like a hill, has a slope.

The slope of a line is given as a number. When you move between two points on the line, the slope is found by dividing the change in y values by the corresponding change in x values.

$$\text{Slope of a line} = \frac{\text{Change in } y \text{ value (vertical change)}}{\text{Change in } x \text{ value (horizontal change)}}$$

- A line that goes up from left to right has a *positive slope*.
- A line that goes down from left to right has a *negative slope*.

In Example 1, the slope of the line is 2.

In Example 2, the slope of the line is $-\left(\frac{2}{2}\right) = -1$.

EXAMPLE 1 $\text{Slope} = \dfrac{4}{2} = 2$

EXAMPLE 2 $\text{Slope} = -\dfrac{2}{2} = -1$

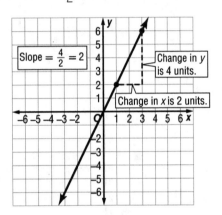

The line above has a *positive slope* because it goes *up* from left to right.

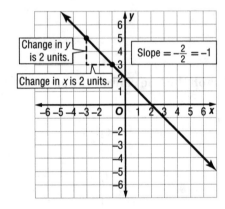

The line above has a *negative slope* because it goes *down* from left to right.

Zero Slope and Undefined Slope

Two types of lines have neither positive nor negative slope.
- A horizontal line has *zero slope,* or the slope equals zero. The x-axis (or any horizontal line) is a line with 0 slope.
- A vertical line has an *undefined slope.* The concept of slope does not apply to a vertical line. The y-axis (or any vertical line) is a line with undefined slope.

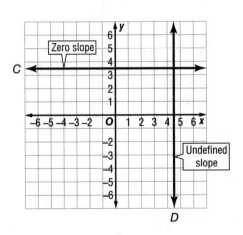

Solve the following problems.

1. Identify the slope of each line as *positive, negative, zero,* or *undefined.*

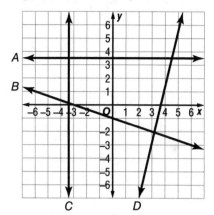

Line *A:* _____   Line *B:* _____

Line *C:* _____   Line *D:* _____

2. Find the numerical value of the slope of line *E.* Identify the *x-* and *y-*intercepts.

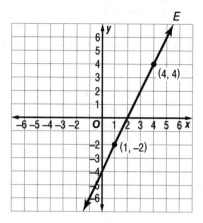

Slope of line *E* = _____

*x-*intercept = (,   )

*y-*intercept = (,   )

3. Find the numerical value of the slope of line *F.* Identify the *x-* and *y-*intercepts.

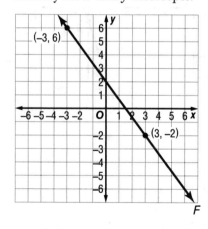

Slope of line *F* = _____

*x-*intercept = (,   )

*y-*intercept = (,   )

4. Subtract the coordinates to find the slope of the line that passes through each pair of points. Part a is done as an example.

   a. (3, 1) and (4, 6)

   $$slope = \frac{change\ in\ y}{change\ in\ x} = \frac{6-1}{4-3} = \frac{5}{1} = 5$$

   b. (0, 2) and (1, 4)

   c. (1, 2) and (−2, 5)

   d. (0, 0) and (3, 2)

Graphing a Linear Equation

When solutions of a linear equation such as $y = 3x - 1$ are plotted on a coordinate grid, they always lie on a straight line. Drawing this line of solutions is called **graphing the equation.**

To graph a linear equation, follow these steps.

STEP 1 Choose three values for x and find the matching values for y. Write each pair of values in a Table of Values.

STEP 2 Plot the three points on a coordinate grid; then connect them with a line extending to the edges of the grid.

Note: Only two points are needed to draw a line, but the third point serves as a check that you plotted the first two points correctly.

EXAMPLE Graph the equation $y = 3x - 1$.

STEP 1 Let $x = 0$, 1, and 2. Solve the equation $y = 3x - 1$ for these three values. Make a Table of Values.

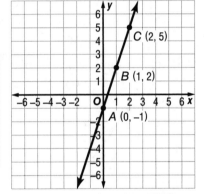

x value	$y = 3x - 1$
$x = 0$	$y = 3(0) - 1 = -1$
$x = 1$	$y = 3(1) - 1 = 2$
$x = 2$	$y = 3(2) - 1 = 5$

Table of Values

x	y
0	−1
1	2
2	5

STEP 2 Plot the points $A = (0, -1)$, $B = (1, 2)$, and $C = (2, 5)$. Connect them with a line that extends to the edges of the grid.

Graph each equation. As a first step, complete the Table of Values. Then find the x- and y-intercepts and slope for each graphed line.

1. $y = x - 2$

Table of Values

x	y
0	
2	
5	

x-intercept =

y-intercept =

slope =

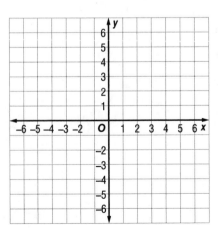

2. $y = x + 2$

Choose your own values for x.

Table of Values	
x	**y**

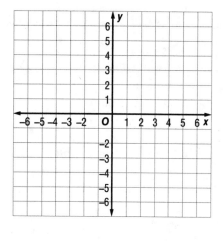

x-intercept =

y-intercept =

slope =

3. $y = -2x - 4$

Choose your own values for x.

Table of Values	
x	**y**

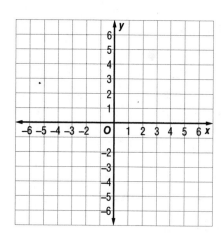

x-intercept =

y-intercept =

slope =

4. $y = \frac{3}{4}x - 1$

Table of Values	
x	**y**
0	
2	
4	

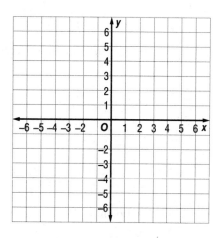

x-intercept =

y-intercept =

slope =

Applying Your Skills: Graphing Circumference

The formula for the circumference of a circle, $C = 2\pi r$, is a linear equation containing the variables C and r. The number π is not a variable. The approximate value of π is 3.14.

Use the grid at the right to answer the problems below.

1. Graph the equation $C = 2\pi r$ for positive values of r and C. For the Table of Values, choose four values for r from among those shown on the grid. Round to the nearest tenth.

Table of Values	
r	**C**

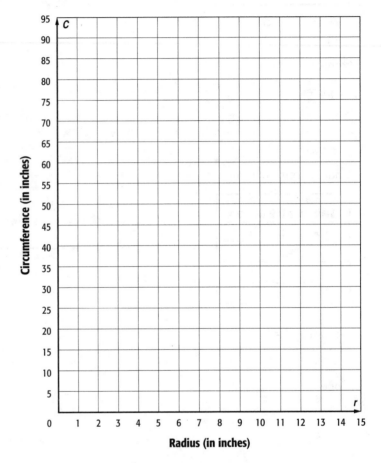

2. Estimate the circumference of a circle that has a radius of $6\frac{1}{2}$ inches.

3. Estimate the radius of a circle that has a circumference of 48 inches.

4. Estimate the slope of the graphed line $C = 2\pi r$.

$$\text{slope} = \frac{\text{change in } C}{\text{change in } r}$$

Applying Your Skills: Graphing Temperature

The formula that relates degress Celsius (°C) to degrees Fahrenheit (°F) is $°C = \frac{5}{9}(°F - 32°)$. This formula is a linear equation containing the variables °F and °C. When finding values, remember to find the difference within parentheses before multiplying by $\frac{5}{9}$.

Use the grid at the right to answer the problems below.

1. Graph the equation $°C = \frac{5}{9}(°F - 32°)$ for positive values of °F.

    For the Table of Values, choose four values for °F from among those shown on the grid.

Table of Values	
°F	**°C**

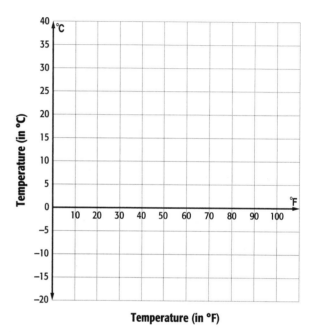

Temperature (in °F)

2. Estimate the temperature in °C when a Fahrenheit thermometer reads 65°F.

3. At what temperature does the graphed line cross the horizontal (°F) axis?

4. Estimate the slope of the graphed line $°C = \frac{5}{9}(°F - 32°)$.

    $$\text{slope} = \frac{\text{change in °C}}{\text{change in °F}}$$

Graphing a System of Equations

On page 100, you learned that two linear equations with a common solution is called a system of equations. One way to find a common solution is to subtract one equation from the other.

A second way to find a common solution is to graph each equation on the same graph. The point at which the two lines cross is the common solution.

EXAMPLE By graphing, find the common solution to these equations.

Equation 1: $y = 2x + 3$
Equation 2: $y = x + 2$

STEP 1 Make a Table of Values for each equation.

Table of Values for $y = 2x + 3$	
x	**y**
−2	−1
0	3
1	5

Table of Values for $y = x + 2$	
x	**y**
0	2
2	4
4	6

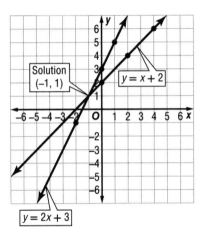

STEP 2 Graph each equation.

STEP 3 Identify the point at which the lines intersect. This point is (−1, 1).

ANSWER: **The common solution is (−1, 1).**

You can check this answer by substituting −1 into the equations for x. When you do this, you should get $y = 1$.

Find a common solution for the two equations by graphing. Complete the Table of Values. Check your answers by substituting the values into the equations.

1. Equation 1: $y = 2x + 1$
   Equation 2: $y = -x + 4$

Table of Values for $y = 2x + 1$	
x	**y**
−2	
0	
2	

Table of Values for $y = -x + 4$	
x	**y**
0	
2	
4	

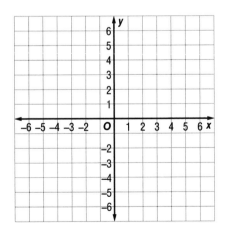

Common solution: (,   )

2. Equation 1: $y = 2x + 3$
Equation 2: $y = x + 1$

Table of Values for $y = 2x + 3$	
x	**y**
−1	
0	
1	

Table of Values for $y = x + 1$	
x	**y**
0	
2	
4	

Common solution: (,   )

3. Equation 1: $y = 3x + 1$
Equation 2: $y = -x + 5$

Table of Values for $y = 3x + 1$	
x	**y**
−1	
0	
1	

Table of Values for $y = -x + 5$	
x	**y**
−1	
0	
6	

Common solution: (,   )

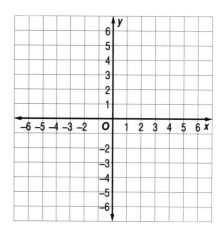

Graphing a Quadratic Equation

When a quadratic equation such as $y = x^2 - 3$ is graphed, the resulting graph is a curve called a **parabola.**

To graph a quadratic equation, follow these steps.

STEP 1 Choose both negative and positive values for x. Then find the matching values for y. Write these in a table.

STEP 2 Plot each point on the grid, and connect them with a curved line. Extend the curve to the edges of the grid.

EXAMPLE 1 Graph the equation $y = x^2 - 3$.

STEP 1 Let $x = -3, -2, -1, 0, 1, 2,$ and 3. Solve the equation $y = x^2 - 3$ for these values. Make a Table of Values.

x value	$y = x^2 - 3$
$x = -3$	$y = (-3)^2 - 3 = 6$
$x = -2$	$y = (-2)^2 - 3 = 1$
$x = -1$	$y = (-1)^2 - 3 = -2$
$x = 0$	$y = (0)^2 - 3 = -3$
$x = 1$	$y = (1)^2 - 3 = -2$
$x = 2$	$y = (2)^2 - 3 = 1$
$x = 3$	$y = (3)^2 - 3 = 6$

Table of Values	
x	**y**
−3	6
−2	1
−1	−2
0	−3
1	−2
2	1
3	6

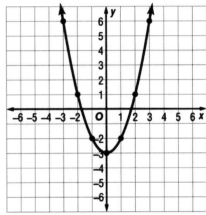

STEP 2 Plot each point and connect them with a smoothly curving line. See the completed graph.

EXAMPLE 2 Use the graph from Example 1 to estimate x when $y = 4$.

STEP 1 Locate $+4$ on the y-axis.

STEP 2 From $+4$ on the y-axis, read the left and right points on the curve. Read down to the x-axis for the two x values.

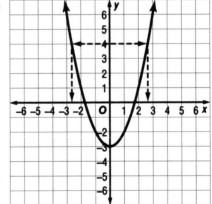

ANSWER: x **can have either of two values:**
$x \approx -2.7$ **and** $x \approx 2.7$

These values are the two approximate square roots of 7. Do you know why?

In the graphed equation, $x^2 = 7$ when $y = 4$.

For each problem, graph the equation. As a first step, complete the Table of Values. (The first value is done for you.) Then use the graph to estimate the two values for x as indicated.

1. **a.** $y = x^2 - 5$

Table of Values	
x	**y**
−4	11
−3	
−2	
0	
2	
3	
4	

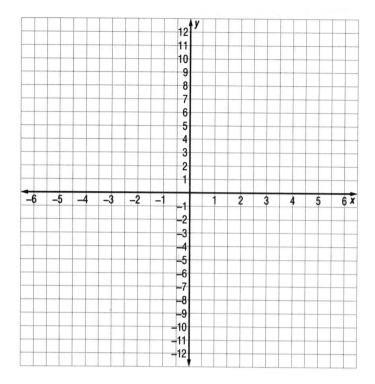

b. Estimate the two values of x when $y = 6$.

2. **a.** $y = 2x^2 - 6$

Table of Values	
x	**y**
−3	12
−2	
−1	
0	
1	
2	
3	

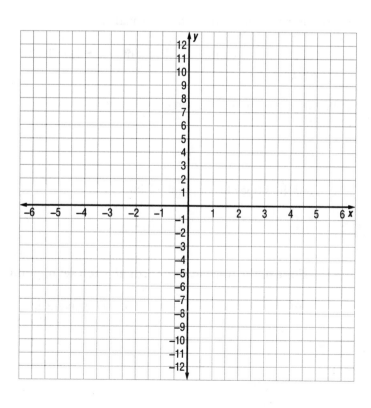

b. Estimate the two values of x when $y = 5$.

Graphing Equations **Review**

Solve the problems below. When you finish, check your answers at the back of the book. Then correct any errors.

For problems 1 and 2, refer to the grid at the right.

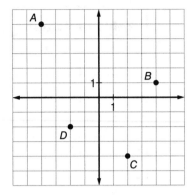

1. On the grid, label the following:

   a. the *x*-axis with an *x*

   b. the *y*-axis with a *y*

   c. the origin with (0, 0)

   d. the *x*-coordinates from −6 to +6

   e. the *y*-coordinates from −6 to +6

2. Write the coordinates of each point.

   Point *A* = (,   )

   Point *B* = (,   )

   Point *C* = (,   )

   Point *D* = (,   )

For problem 3, refer to the grid at the right.

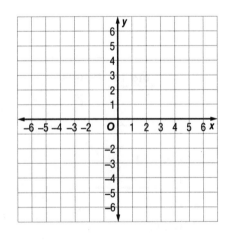

3. **a.** Graph the line containing the points (−6, −4), (−3, −3), and (3, −1).

   b. Identify the *x*- and *y*-intercepts of the line you drew.

   x-intercept = (,   )

   y-intercept = (,   )

   c. Find the slope of your graphed line.

   slope =

Find the slope of the line passing through each set of points.

4. **a.** (0, 0) and (2, 4)   **b.** (1, 0) and (2, 5)   **c.** (2, 3) and (4, 3)

5. Graph the equation $y = 2x - 4$. Choose your own values for x. Find the x- and y-intercepts and the slope of the graphed line.

Table of Values	
x	**y**

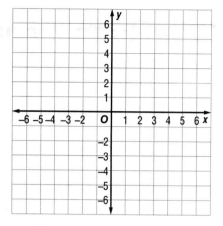

x-intercept =

y-intercept =

slope =

6. Find the common solution for the two equations by graphing.

$y = 2x - 1$

$y = x + 2$

Common solution: (,   )

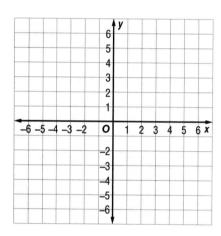

7. Graph the equation $y = x^2 - 6$ by first filling in the Table of Values. The first value is done for you.

Table of Values	
x	**y**
−4	10
−3	
−2	
0	
2	
3	
4	

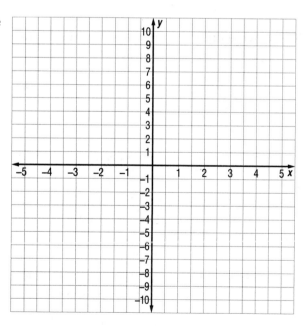

What Is an Inequality?

In an earlier chapter, you learned that an equation is a statement that two quantities are equal. In this chapter, you will learn about inequalities. An **inequality** is a statement that two quantities are *not* equal.

Five comparison symbols are used with inequalities.

Symbol	Meaning	Example	Read as
$<$	is less than	$b < 9$	b is less than 9
$>$	is greater than	$n > 16$	n is greater than 16
$\leq$	is less than or equal to	$x \leq 5$	x is less than or equal to 5
$\geq$	is greater than or equal to	$s \geq -4$	s is greater than or equal to -4
$\neq$	is not equal to	$c \neq 7$	c is not equal to 7

In an inequality, the variable can be *any value* for which the inequality is true.

- $b < 9$: b can be any value less than 9.
  Example values: $b = -2.5$, $b = 0$, or $b = 8$
- $n > 16$: n can be any value greater than 16.
  Example values: $n = 16.3$, $n = 17$, or $n = 31\frac{1}{4}$
- $a \leq 7$: a can be any value less than or equal to 7.
  Example values: $a = -3.5$, $a = 1$, or $a = 7$
- $r \geq -2$: can be any value greater than or equal to -2.
  Example values: $r = -2$, $r = 0$, or $r = 2\frac{1}{2}$

Did You Know ...?

In an inequality, the allowed values are whole numbers, decimals, and fractions.

**For each inequality, circle any allowed value for the variable.
Two problems are done as examples.**

1. a. $m < 7$ ⊝3⊝   ⊙0⊙   ⊙5⊙   7   9   14   **b.** $n < 10$ -2 0   3   7   10   15

2. a. $s > 6$ -6 -2 6   8   10   19   **b.** $y > 3$ -3 0   3   6   11   23

3. a. $n \leq 12$ ⊝-8⊝   ④4④   ⑫12⑫   13   19   25   **b.** $p \leq 6$ -5 0   3   5   6   27
   [n can be any number less than or equal to 12.]   [p can be any number less than or equal to 6.]

4. a. $x \geq 14$ -6 0   2   17   29   41   **b.** $z \geq -4$ -9 -5 -4 -3 0   4
   [x can be any number greater than or equal to 14.]   [z can be any number greater than or equal to -4.]

Graphing an Inequality

The values for which an inequality is true can be graphed on a number line.
- A solid circle ● means that the number *is* an allowed value.
- An open circle ○ means that a number is *not* an allowed value.
- A solid line is drawn through all allowed values.

EXAMPLE 1 Inequality: $n > -5$

Graph:

Values of n

$$-6\ -5\ -4\ -3\ -2\ -1\ \ 0\ \ 1\ \ 2\ \ 3\ \ 4\ \ 5\ \ 6$$

Meaning: n is greater than -5 (n *cannot* have the value -5, indicated by ○.)

EXAMPLE 2 Inequality: $r \geq 0$

Graph:

Values of r

$$-6\ -5\ -4\ -3\ -2\ -1\ \ 0\ \ 1\ \ 2\ \ 3\ \ 4\ \ 5\ \ 6$$

Meaning: r is greater than or equal to 0. (r *can* have the value 0, indicated by ●.)

EXAMPLE 3 Inequality: $y \leq -2$

Graph:

Values of y

$$-6\ -5\ -4\ -3\ -2\ -1\ \ 0\ \ 1\ \ 2\ \ 3\ \ 4\ \ 5\ \ 6$$

Meaning: y is less than or equal to -2. (y *can* have the value -2, indicated by ●.)

Circle the letter for the inequality graphed on each number line.

1.

Values of s

$$-6\ -5\ -4\ -3\ -2\ -1\ \ 0\ \ 1\ \ 2\ \ 3\ \ 4\ \ 5\ \ 6$$

  a. $s < -3$ **b.** $s > -3$ **c.** $s \geq -3$

2.

Values of r

$$-6\ -5\ -4\ -3\ -2\ -1\ \ 0\ \ 1\ \ 2\ \ 3\ \ 4\ \ 5\ \ 6$$

  a. $r < 1$ **b.** $r \leq 1$ **c.** $r \geq 1$

Write the inequality graphed on each number line.

3.

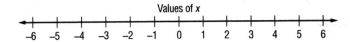

4.

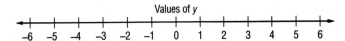

Graph each inequality.

5. $x > -2$

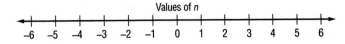

6. $y \leq 4$

Values of y

7. $n \geq 0$

Values of n

8. $s < 3$

Values of s

Solving an Inequality

In an inequality, the variable may not be standing alone and the allowed values may not be easy to see. For example, look at the inequality $2x + 4 \geq 5$. To find the allowed values of x, you solve this inequality in a way similar to the way you solve an equation.

To solve an inequality is to write it so that the variable stands alone.

To get the variable standing alone, perform the same operations on each side of the inequality sign.

- Add (or subtract) the same number to (or from) each side.
  (See Example 1.)
- Multiply or divide each side by the same number.
  (See Example 2.)
- Use two or more operations, starting with addition or subtraction.
  (See Examples 3 and 4.)

EXAMPLE 1 Solve for x: $x - 5 \leq 2$

> **STEP 1** Add 5 to each side of the inequality.
>
> $x - 5 + 5 \leq 2 + 5$
>
> **STEP 2** Simplify each side of the inequality.

ANSWER: $x \leq 7$

EXAMPLE 2 Solve for y: $4y \geq 16$

> **STEP 1** Divide each side of the inequality by 4.
>
> $\dfrac{4y}{4} \geq \dfrac{16}{4}$
>
> **STEP 2** Simplify each side of the inequality.

ANSWER: $y \geq 4$

EXAMPLE 3 Solve for y: $5y - 3 \leq 27$

> **STEP 1** Add 3 to each side of the inequality.
>
> $5y - 3 + 3 \leq 27 + 3$
>
> **STEP 2** Simplify each side of the inequality.
>
> $5y \leq 30$
>
> **STEP 3** Divide each side of the inequality by 5.
>
> $\dfrac{5y}{5} \leq \dfrac{30}{5}$
>
> **STEP 4** Simplify each side of the inequality.

ANSWER: $y \leq 6$

EXAMPLE 4 Solve for n: $\dfrac{n}{2} + 6 \leq 13$

> **STEP 1** Subtract 6 from each side of the inequality.
>
> $\dfrac{n}{2} + 6 - 6 \leq 13 - 6$
>
> **STEP 2** Simplify each side of the inequality.
>
> $\dfrac{n}{2} \leq 7$
>
> **STEP 3** Multiply each side of the inequality by 2.
>
> $2\left(\dfrac{n}{2}\right) \leq 2(7)$
>
> **STEP 4** Simplify each side of the inequality.

ANSWER: $n \leq 14$

Solve each inequality.

1. $2x + 6 \geq 10$ 　　　　　$3n - 4 \geq 11$ 　　　　　$5z + 5 > 25$

2. $6w + 2 \leq 14$ 　　　　　$4x - 1 < -9$ 　　　　　$2y + 4 \leq 0$

3. $\frac{n}{2} + 3 > 9$ 　　　　　$\frac{x}{5} - 6 \geq 2$ 　　　　　$\frac{c}{4} - 3 < -2$

4. $\frac{2}{3}x + 1 \leq 5$ 　　　　　$\frac{3}{4}n + 3 < -3$ 　　　　　$\frac{5}{8}y - 4 \leq 21$

If you multiply or divide each side of an inequality by −1, you need to reverse the inequality sign.

EXAMPLE 5　**Solve for y: $-y \leq 6$**

STEP 1　Multiply each side of the inequality by −1 and change the inequality sign from ≤ to ≥.

$-1(-y) \geq -1(6)$

STEP 2　Simplify both sides of the inequality.

ANSWER: $y \geq -6$

EXAMPLE 6　**Solve for x: $3 - x > 5$**

STEP 1　Subtract 3 from each side of the inequality and simplify.

$3 - 3 - x > 5 - 3$
$-x > 2$

STEP 2　Multiply each side of the inequality by −1, change > to <, and simplify.

$-1(-x) < -1(2)$
$x < -2$

ANSWER: $x < -2$

Solve each inequality.

5. $-x \geq 8$ 　　　　　$4 - n < 10$ 　　　　　$6 - 2x > 10$

Solve and graph the following inequalities.

6. $3x + 7 > 10$

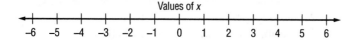

7. $4n + 5 \leq 1$

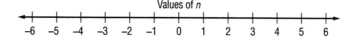

8. $7y + 2 > 2$

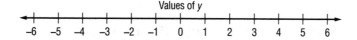

9. $3 - x > 4$

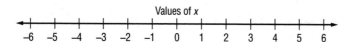

10. $-2y + 5 < 1$

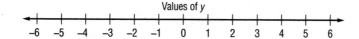

Applying Your Skills: Inequalities

Circle the correct inequality and solve for the allowed values of the variable.

1. If 6 is added to a number *n*, the sum is greater than 10. Which inequality tells the possible values of *n*?

   a. $n + 6 > 4$
   b. $n - 10 > 6$
   c. $n + 6 > 10$

   Solution:

2. If 8 is subtracted from 3 times a number *m*, the difference is less than or equal to 4. Which inequality tells the possible values of *m*?

   a. $3m - 8 \leq 4$
   b. $8 - 3m \leq 4$
   c. $3m - 8 \geq 4$

   Solution:

Solve.

3. Suppose you are working Saturday at a poster shop. You will be paid $75 plus an extra $1.50 for each large poster you sell. Your goal is to earn at least $125 on Saturday.

   a. Write an inequality that tells the number *n* of posters you must sell on Saturday to reach or exceed your goal.

   b. What is the least number of posters you can sell in order to earn $125 or more?

4. The perimeter of $\triangle ABC$ is greater than 75 yards.

   a. Write an inequality that describes the possible values for the length *c* of the third side.

   b. To the nearest yard, what is the smallest value *c* may have?

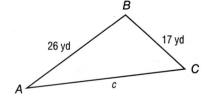

5. You have been hired to design a garden that is to be 54 feet long and no larger in area than 1,800 square feet.

   a. Write an inequality that describes the range of possible values for the width *w* of the garden.

   b. To the nearest foot, what is the largest value *w* may have?

Two-Part Inequalities

A two-part inequality contains two comparison symbols that are used to describe *a range of allowed values*. The two types of two-part inequalities are *and* **inequalities** and *or* **inequalities**.

And Inequalities

Inequalities whose allowed values run *between* two numbers are called *and inequalities*. The allowed values of an *and inequality* can be graphed as a single section of a number line.

Examples	Meaning	Graphed Values
$2 < n < 6$	n is greater than 2 *and* n is less than 6.	
$-3 \leq m < 4$	m is greater than or equal to -3 *and* m is less than 4.	
$-5 \leq y \leq 5$	y is greater than or equal to -5 *and* y is less than or equal to 5.	

Or Inequalities

In an *or inequality*, allowed values are graphed as separate sections of a number line.

Examples	Meaning	Graphed Values
$s < 2$ or $s \geq 4$	s is less than 2 *or* s is greater than or equal to 4.	
$x \leq -1$ or $x > 3$	x is less than or equal to -1 *or* x is greater than 3.	

Write the meaning of each inequality.

1. $-5 < n < 8$ _____

2. $-1 \leq x < 11$ _____

3. $n \leq 0$ or $n > 7$ _____

4. $p < -3$ or $p \geq 5$ _____

For each inequality, three values are given. Circle Yes if the value is a possible value for the inequality, No if it is not.

5. $-2 < y < 5$ **a.** $y = 3$ Yes No   **b.** $y = -2$ Yes No   **c.** $y = 0$ Yes No

6. $-5 \leq x < 0$ **a.** $x = 0.6$ Yes No   **b.** $x = 5$ Yes No   **c.** $x = -5$ Yes No

7. $2 < n \leq 8$ **a.** $n = -3$ Yes No   **b.** $n = 5\frac{1}{3}$ Yes No   **c.** $n = 8$ Yes No

8. $y \leq 3$ or $y > 8$ **a.** $y = -1.5$ Yes No   **b.** $y = 0$ Yes No   **c.** $y = 8$ Yes No

9. $x < -2$ or $x \geq 3$ **a.** $x = -4$ Yes No   **b.** $x = 1$ Yes No   **c.** $x = 3$ Yes No

10. $n \leq 1$ or $n \geq 5$ **a.** $n = -1$ Yes No   **b.** $n = 0$ Yes No   **c.** $n = 5\frac{3}{4}$ Yes No

Write the inequality graphed on each number line.

11.

Values of x

12.
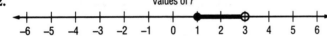
Values of r

13.
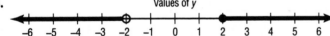
Values of y

14.
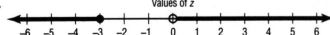
Values of z

Applying Your Skills: Two-Part Inequalities

Graph each inequality.

1. $-3 \leq n < 4$

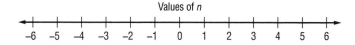

2. $x < -1$ or $x \geq 4$

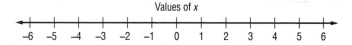

Solve.

3. During the year, Mark's weight varies from 172 pounds to 178 pounds. He tends to gain weight during the winter and reduce during the summer. Write an inequality that describes the range of Marks weight w.

4. Each day, 5% to 10% of the 2,000 students who attend Hoover High School are absent. Write an inequality that describes the number n of students who are absent on an average day at Hoover High School.

5. The perimeter of $\triangle DEF$ is greater than 102 inches but less than 119 inches.

    a. Write an inequality that describes the range of possible values for the length c of the third side.

    b. If c is a whole number, what is the smallest value that c may be?

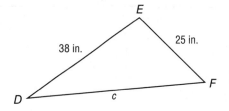

6. Geraldine conducted a survey. She contacted only people who have been driving for less than 1 year or for more than 5 years. Write an inequality that gives the range of values for the length of time t that the people in the survey have been driving.

Inequalities Review

Solve each problem below. When you finish, check your answers at the back of the book. Then correct any errors.

Write the meaning of each inequality.

1. $x < 6$ _____

2. $n > 9$ _____

3. $y \geq -5$ _____

4. $s \leq 4$ _____

Write the inequality graphed on each number line.

5.

Values of p

6.
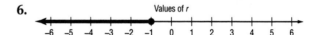
Values of r

7.
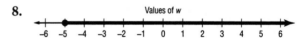
Values of n

8.
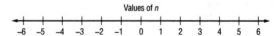
Values of w

Graph each inequality.

9. $x \leq 5$

Values of x

10. $n > -3$

Values of n

Solve each inequality.

11. $3x + 4 \leq 10$

12. $2n - 5 \geq 15$

13. $2n + 5 < 45$ **14.** $2x + 7 > 23$

Write the meaning of each inequality.

15. $-2 < x < 7$ _____

16. $0 \leq y < 6$ _____

17. $n \leq -2$ or $n \geq 5$ _____

18. $r < 2$ or $r > 7$ _____

Write the inequality graphed on each number line.

19.

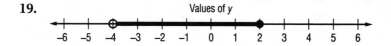

20.

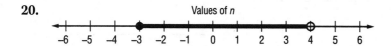

Solve.

21. The perimeter of $\triangle ABC$ is less than 94 centimeters but greater than 85 centimeters.

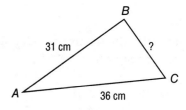

a. Write an inequality that describes the range of possible values for the length of side BC.

b. If side BC is a whole number, what is the shortest length that side BC may be?

POLYNOMIALS

Naming Polynomials

A **polynomial** is an algebraic expression that contains one or more terms combined by addition. Each term is the product of a number (coefficient) and one or more variables. Each variable may have an exponent. A number standing alone is also a term.

The three most common types of polynomials are **monomials, binomials,** and **trinomials.**

- A monomial has one term.
- A binomial has two terms.
- A trinomial has three terms.

Monomials	**Binomials**	**Trinomials**
z	$x - 3$	$x + y + z$
$-2x$	$3y + 9$	$5r - 3s + 6t$
$3a^2$	$-5c^2 + 2c$	$4y^2 - 5y + 7$
$-6x^2y^2$	$4a^2b^2 + 2ab$	$7y^3 - 10y^2 + 9y$

To identify a term, change a subtraction sign to an addition sign and change the sign of the coefficient of the term. (Recall that subtraction is equivalent to adding a negative number.) For example, in the binomial $x - 3$ the terms are x and -3. In the trinomial $7y^3 - 10y^2 + 9y$, the terms are $7y^3$, $-10y^2$, and $9y$.

Identify each polynomial as a monomial, binomial, or trinomial.

1. $5x - 6$ _____

2. $4x + 6y - z$ _____

3. $-y$ _____

4. $-5y^2 + 3y$ _____

5. $r^2 + s^2 - t^2$ _____

6. $7x^2y^2$ _____

Identify the terms in each polynomial. The first problem in each row is done as an example.

7. $-12x + 7$
  $-12x$ and 7

$9d + 6$

$-3a^2 + 5a$

$4y + \frac{1}{2}$

8. $9a - 2b$
  $9a$ and $-2b$

$-6d - 8e$

$8x^2 - 3y^2$

$\frac{2}{3}z^2 - 7z$

9. $2x^2 + 7x - 4$
  $2x^2, 7x,$ and -4

$3y^2 - 5y - 8$

$8x - 4y + 13$

Like Terms in Polynomials

Like terms in a polynomial are terms that have the same variables raised to the same exponents. A term includes its coefficient and sign.

EXAMPLE Like terms are circled in the pairs of polynomials below.

$2x^3 + \boxed{3x^2} + 7$

⟵ like terms

$\boxed{5x^2} + 2x$

$6x^4 + x^2 \boxed{-5x}$

⟵ like terms

$x^3 + \boxed{4x} - 1$

Only like terms can be combined when polynomials are added or subtracted.

Evaluating Polynomials

To evaluate a polynomial is to find its value for a particular value of each variable. A polynomial is evaluated just like any other algebraic expression.

EXAMPLE To find the value of $x^3 + 2x^2 - 4$ for $x = 3$, substitute and simplify.

$$x^3 + 2x^2 - 4 = 3^3 + 2(3)^2 - 4 \qquad \text{Substitute 3 in place of } x.$$
$$= 27 + 2(9) - 4 \qquad \text{Find each power.}$$
$$= 27 + 18 - 4 \qquad \text{Simplify.}$$
$$= \mathbf{41}$$

ANSWER: For $x = 3$, $x^3 + 2x^2 - 4$ has value 41.

Circle any like terms in each pair of polynomials.

1. $x^2 + 3x$ and $x^3 - 5x$ $\qquad$ $2y^3 - y$ and $4y^3 - 2y^2$ $\qquad$ $-2z^4 + 2z^3$ and $z^2 - z$

2. $2x^3 - 5$ and $x^2 - 9$ $\qquad$ $n^3 + 2n$ and $3n^2 - n$ $\qquad$ $5x^4 + 4x^3$ and $x^3 - x$

3. $y^5 + 2y$ and $2y^3 + 4y^2$ $\qquad$ $3r^3 - r$ and $4r^2 + 3r$ $\qquad$ $-3a^4 + 2a^2$ and $a^4 - a$

Evaluate each polynomial for the value of the variable given.

4. $2x^4 + x^3 - x^2$ for $x = 2$ $\qquad$ 5. $3n^3 + 2n^2 - 8$ for $n = 0$

6. $5y^3 - 7y^2 + y + 1$ for $y = 1$ $\qquad$ 7. $-x^3 + 2x^2 + 3x + 5$ for $x = 3$

Adding Polynomials

To add polynomials, add the coefficients of like terms. Follow the rules for adding signed numbers.

EXAMPLE 1 Add: $3x$, $2x$, and $4y$

$3x + 2x + 4y = \mathbf{5x + 4y}$
$(3 + 2 = 5)$
x and y are not like terms.

EXAMPLE 2 Add: $-5a^2b$ and a^2b

$-5a^2b + a^2b = \mathbf{-4a^2b}$
$(-5 + 1 = -4)$
The coefficient of a^2b is -4.

EXAMPLE 3 Add: $(6x - 9y) + (-5x + 4y)$

STEP 1 Remove the parentheses from each binomial.
$6x - 9y + (-5x) + 4y$

STEP 2 Combine like terms.
x terms: $6x + (-5x) = \mathbf{x}$ $\qquad (6 + -5 = 1)$
y terms: $-9y + 4y = \mathbf{-5y}$ $\qquad (-9 + 4 = -5)$

ANSWER: $x + -5y$ or $x - 5y$

(**Remember:** An added negative number can be written as a subtracted positive number.)
The answer can be written either way. However, $x - 5y$ is the usual way to write this answer.

Add. The first problem in each row is done as an example.

Adding Monomials

1. $6y + 3y = \mathbf{9y}$ $\qquad\qquad$ $5r + 2r =$ $\qquad\qquad$ $12s + 7s =$ $\qquad\qquad$ $10x + 5x =$

2. $4z + z = \mathbf{5z}$ $\qquad\qquad$ $7a + a =$ $\qquad\qquad$ $2c + c =$ $\qquad\qquad$ $11n + n =$

3. $8x + (-3x) = \mathbf{5x}$ $\qquad\qquad$ $9y + (-2y) =$ $\qquad\qquad$ $15r + (-4r) =$ $\qquad\qquad$ $7y + (-7y) =$

4. $2x^2 + (-4x^2) = \mathbf{-2x^2}$ $\qquad$ $6y^2 + (-y^2) =$ $\qquad\qquad$ $7z^3 + (-2z^3) =$ $\qquad\qquad$ $12r^3 + (-4r^3) =$

Adding Binomials and Monomials

5. $(3y + 5) + 7 = 3y + 12$ $(4x + 9) + 4 =$ $(7z + 6) + (-3) =$

6. $(4z + 2) + -5z = -z + 2$ $(7y + -8) + y =$ $(12s + 4) + 4s =$

7. $(7n^2 + n) + 9n^2 = 16n^2 + n$ $(6x^2 + 5x) + 3x =$ $(3r^2 + r) + (-2r^2) =$

Adding Binomials and Binomials

8. $(2z + 4) + (-3z + 5) = -z + 9$

$(4x + 6) + (-2x + 5) =$

$(5n + 1) + (-n + 7) =$

9. $(5a + 3b) + (-2a - 8b) = 3a - 5b$

$(2x + 9y) + (-x - 5y) =$

$(8r + 3s) + (-3r - 4s) =$

10. $(4z^2 + 6) + (-z^2 - 4) = 3z^2 + 2$

$(6n^2 + 5) + (-3n^2 - 2) =$

$(2y^2 + 9y) + (-y^2 - 5y) =$

11. $(5a^2 + 3b) + (a^2 - 5b) = 6a^2 - 2b$

$(6r^2 + 5s) + (2r^2 - 8s) =$

$(9m^2 + n) + (4m^2 - 3n) =$

Write the perimeter of each figure as a trinomial.

12.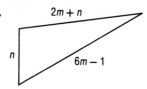

13.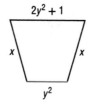

14.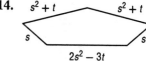

Subtracting Polynomials

To subtract polynomials, subtract the coefficients of like terms. To subtract coefficients, follow the rules for subtracting signed numbers.
- Change the subtraction sign (−) to an addition sign (+).
- Change the sign of each term being subtracted.

EXAMPLE 1   Subtract: $9y - 4y$

Subtract the coefficients of the y terms.

$9 - 4 = \mathbf{5}$ or $9 + (-4) = \mathbf{5}$

ANSWER: **5y**

EXAMPLE 2   Subtract: $5c^2 - (-6c^2)$

Subtract the coefficients of the c^2 terms.

$5 - (-6) = 5 + 6 = \mathbf{11}$

ANSWER: **11c²**

(**Remember:** To subtract, change − to + and change −6 to 6.)

When a subtraction sign precedes a binomial in parentheses,
- remove the parentheses by distributing the −1 to each term, and
- subtract each term that was within the parentheses.

EXAMPLE 3   Subtract: $(5x - y) - (3x + 4y)$

STEP 1   Remove the parentheses from each binomial.

$(5x - y) = 5x - y$
$-(3x + 4y) = -3x - 4y$

Subtract each term. ⎯⎯⎯

STEP 2   Combine like terms.

x terms: $5x - 3x = \mathbf{2x}$
y terms: $-y - 4y = \mathbf{-5y}$

ANSWER: **2x − 5y**

Subtract. The first problem in each row is done as an example.

1. $5x - 2x = \mathbf{3x}$ $7z - 2z =$ $10r - 6r =$ $15x - 8x =$

2. $6z - (-3z) = \mathbf{6z + 3z}$ $9b - (-b) =$ $6d - (-2d) =$ $21n - (-17n) =$
    $= \mathbf{9z}$

3. $(7x + 2y) - 5x = $ **$7x - 5x + 2y = 2x + 2y$**   **4.** $(3x - 3y) - 5y = $ **$3x - 3y - 5y = 3x - 8y$**

$(8m + 3n) - 2m =$ $(9a + 5b) - 6a =$

$(5r - 6s) - 7s =$ $(12c - 3d) - 7d =$

5. $(8a - 3b) - (2a + b) = $ **$8a - 2a - 3b - b = 6a - 4b$**

$(9x - 2y) - (6x + 2y) =$

$(12m + 3n) - (5m + 5n) =$

6. $(5x^2 + 4x - 7) - (2x^2 + 2) = $ **$5x^2 - 2x^2 + 4x - 7 - 2 = 3x^2 + 4x - 9$**

$(9a^2 + 2a - 8) - (3a^2 + 1) =$

$(8n^2 + 7n - 2) - (n^2 + 9n + 2) =$

Write the unlabeled length as a polynomial.

7.

8.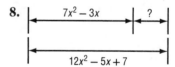

Solve.

9. The perimeter of the figure at the right is given as the polynomial $8n^2 + 4n + 5$. Write the length of the unlabeled side as a polynomial.

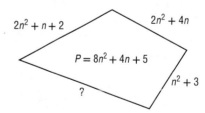

Multiplying Monomials

To multiply monomials, follow these two steps.

STEP 1 Multiply the coefficients according to the rules for multiplying signed numbers.

STEP 2 Multiply the variables by adding the exponents of each like variable. Write the variables in alphabetical order.

(**Note:** To review the rules for multiplying like bases by adding exponents, see page 30.)

Rule for multiplying powers of like variables:
Add the exponents and keep the same variable.

EXAMPLE 1 Multiply: $3x^2$ by $-4x$

STEP 1 Multiply the coefficients.

$3(-4) = \mathbf{-12}$

STEP 2 Add the exponents of the x variables.

$x^2 x = x^2 x^1 = x^{2+1} = \mathbf{x^3}$

ANSWER: $\mathbf{-12x^3}$

EXAMPLE 2 Multiply: $-2a^2b(-3ab^3)$

STEP 1 Multiply the coefficients.

$-2(-3) = \mathbf{6}$

STEP 2 Add the exponents of the a variables and the b variables.

$a^2 a = a^2 a^1 = a^{2+1} = \mathbf{a^3}$
$bb^3 = b^1 b^3 = b^{1+3} = \mathbf{b^4}$

ANSWER: $\mathbf{6a^3b^4}$

(**Remember:** If no exponent is written, it is understood to be 1. In Example 1, $x = x^1$. In Example 2, $a = a^1$ and $b = b^1$.)

Multiply. The first problem in each row is done as an example.

1. $5y^2(3y^3) = \mathbf{15y^5}$ $6x^4(2x^2) =$ $7n(3n^4) =$ $8a^3(6a^4) =$

2. $5x^2 \cdot 5x^2 = \mathbf{25x^4}$ $6n^4 \cdot 6n^4 =$ $4z^6 \cdot 4z^6 =$ $-8x^3 \cdot -8x^3 =$

3. $3x^2(-2x) = \mathbf{-6x^3}$ $-4c^5(5c^3) =$ $9b^4(-7b) =$ $-5s(9s^3) =$

4. $7x^2y(-xy) = \mathbf{-7x^3y^2}$ $-5c^3d^2(cd^2) =$ $3ab^4(-4a^3b) =$ $-8r^2s(-7rs^3) =$

Multiplying a Polynomial by a Monomial

To multiply a binomial or trinomial by a monomial, multiply each term and combine the separate products. Remove parentheses around a binomial or trinomial by multiplying each term within parentheses by the monomial.

EXAMPLE 1 Multiply: $3x(2x^2 + 4)$

STEP 1 Multiply each term of the binomial by $3x$. Keep the + sign that is between the terms in the binomial.

$3x(2x^2 + 4) =$
$3x \cdot 2x^2 + 3x \cdot 4$

STEP 2 Multiply each pair of terms. Add the exponents of x.

$3x \cdot 2x^2 + 3x \cdot 4 =$
$\mathbf{6x^3 + 12x}$

ANSWER: $6x^3 + 12x$

EXAMPLE 2 Multiply: $2a(5a^2 - 4ab + b^2)$

STEP 1 Multiply each term of the trinomial by $2a$. Keep the signs that are between the terms in the trinomial.

$2a(5a^2 - 4ab + b^2) =$
$2a \cdot 5a^2 - 2a \cdot 4ab + 2a \cdot b^2$

STEP 2 Multiply each pair or terms. Add the exponents.

$2a \cdot 5a^2 - 2a \cdot 4ab + 2a \cdot b^2 =$
$\mathbf{10a^3 - 8a^2b + 2ab^2}$

ANSWER: $10a^3 - 8a^2b + 2ab^2$

Multiply. The first problem in each row is done as an example.

1. $4n(3n^2 + 2) = \mathbf{12n^3 + 8n}$

   $5c(c + 7) =$

   $6a(2a^3 - 5) =$

2. $3x^2(2x^2 + 4x) = \mathbf{6x^4 + 12x^3}$

   $2y(4y^2 - 3y) =$

   $3r(3r^3 + 8r^2) =$

3. $ab(3a^2b^2 - 6) = \mathbf{3a^3b^3 - 6ab}$

   $cd(5c^3d + 3) =$

   $xy(x^2y^3 - 2) =$

4. $4mn(-2m^3n^2 - mn) = \mathbf{-8m^4n^3 - 4m^2n^2}$

   $3a^2b(a^2b - 6ab) =$

   $5cd^2(3c^2d - 2cd) =$

5. $3x(4x^3 - 3xy + xy^2) = \mathbf{12x^4 - 9x^2y + 3x^2y^2}$

   $4a(4a^2 + 6ab^2 - 7b^3) =$

   $6cd(c^3d^3 - 2c^2d^2 + cd) =$

Multiplying a Binomial by a Binomial

To multiply a binomial by a binomial follow these steps.

STEP 1 Multiply the first term of the first binomial by each term of the second binomial.

STEP 2 Multiply the second term of the first binomial by each term of the second binomial.

STEP 3 Combine the separate products.

EXAMPLE 1 Multiply: $(2x + 3y)(4x - 5y)$

STEP 1   STEP 1

$$(2x + 3y)(4x - 5y) = 2x(4x - 5y) + 3y(4x - 5y)$$

STEP 2   STEP 2

STEP 1 Multiply $2x$ by $(4x - 5y)$: $2x(4x - 5y) = 2x \cdot 4x + 2x \cdot -5y$

$$= 8x^2 - 10xy$$

STEP 2 Multiply $3y$ by $(4x - 5y)$: $3y(4x - 5y) = 3y \cdot 4x + 3y \cdot -5y$

$$= 12xy - 15y^2$$

STEP 3 Combine: $8x^2 - 10xy + 12xy - 15y^2 = 8x^2 + 2xy - 15y^2$

ANSWER: $8x^2 + 2xy - 15y^2$

Multiply. The first problem in each group is done as an example.

1. $(2y + 3)(4y + 2) = 2y \cdot 4y + 2y \cdot 2 + 3 \cdot 4y + 3 \cdot 2 = 8y^2 + 4y + 12y + 6$
   $$= 8y^2 + 16y + 6$$

   $(4a + 2)(2a + 1)$

   $(5z + 3)(3z + 2)$

2. $(4x + 3y)(2x - y) = 4x \cdot 2x - 4x \cdot y + 3y \cdot 2x - 3y \cdot y = 8x^2 - 4xy + 6xy - 3y^2$
   $$= 8x^2 + 2xy - 3y^2$$

   $(2a + 3b)(3a - 2b)$

   $(3r + 2s)(r - 4s)$

An interesting result occurs when you multiply the sum of two variables by the difference of the same variables:

$$(a + b)(a - b) = a^2 - ab + ab - b^2 = \boldsymbol{a^2 - b^2}$$

The product is the difference of the squares of the variables. In this product, a number can be substituted for either variable, *a* or *b*.

EXAMPLE 2 Multiply: $(x + 3)(x - 3)$

Multiply $x \times x$ and $3 \times (-3)$.

$(x + 3)(x - 3) = \boldsymbol{x^2 - 9}$

ANSWER: $\boldsymbol{x^2 - 9}$

EXAMPLE 3 Multiply: $(5 + n)(5 - n)$

Multiply 5×5 and $n \times (-n)$.

$(5 + n)(5 - n) = \boldsymbol{25 - n^2}$

ANSWER: $\boldsymbol{25 - n^2}$

3. $(a + 2)(a - 2) = \boldsymbol{a^2 - 4}$

$(c + 5)(c - 5) =$

$(y + 8)(y - 8) =$

4. $(6 + x)(6 - x) = \boldsymbol{36 - x^2}$

$(9 + b)(9 - b) =$

$(4 + m)(4 - m) =$

Write the area of each rectangle as a polynomial. The area formula is on page 224.

5.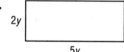
2y

5y

6.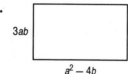
3ab

$a^2 - 4b$

7.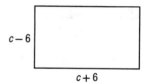
$c - 6$

$c + 6$

Write the volume of each figure as a monomial or a binomial. Formulas are on page 225.

8.
4y

4y

4y

9.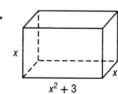
x

x

$x^2 + 3$

10.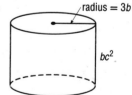
radius $= 3b$

bc^2

Dividing Monomials

To divide monomials, follow these two steps.

STEP 1 Divide the coefficients according to the rules for dividing signed numbers.

STEP 2 Divide the variables by subtracting the exponents of each like variable. Write the variables in alphabetical order.

(**Note:** To review the rules for dividing like bases by subtracting exponents, see page 31.)

Rule for dividing powers of like variables:
Subtract the exponents and keep the same base.

EXAMPLE 1 Divide: $8y^4$ by $-4y$

STEP 1 Divide the coefficients.
$$\frac{8}{-4} = -2$$

STEP 2 Subtract the exponents of the y variables.
$$\frac{y^4}{y} = \frac{y^4}{y^1} = y^{4-1} = y^3$$

ANSWER: $-2y^3$

EXAMPLE 2 Divide: $\dfrac{-16c^5d^2}{-8c^2d}$

STEP 1 Divide the coefficients.
$$\frac{-16}{-8} = 2$$

STEP 2 Subtract the exponents of the c variables and the d variables.
$$\frac{c^5}{c^2} = c^{5-2} = c^3$$
$$\frac{d^2}{d} = \frac{d^2}{d^1} = d^{2-1} = d^1 = d$$

ANSWER: $2c^3d$

(**Remember:** If no exponent is written, it is understood to be 1. In Example 1, $y = y^1$. In Example 2, $d = d^1$. Also, d^1 is written as d in the answer.)

Divide. The first problem in each row is done as an example. (Note: An arrow is used to remind you that you are simplifying an expression.)

1. $\dfrac{y^7}{y^3} \rightarrow y^4$ $\dfrac{x^5}{x^2}$ $\dfrac{a^4}{a^3}$ $\dfrac{c^5}{c^4}$

2. $\dfrac{6b^6}{2b^3} \rightarrow 3b^3$ $\dfrac{9n^3}{-3n^2}$ $\dfrac{-12x^5}{3x^2}$ $\dfrac{-42d^3}{6d^4}$

3. $\dfrac{a^6b^3}{a^4b^2} \rightarrow a^2b$ $\dfrac{m^5n^6}{m^2n^4}$ $\dfrac{-36c^5d^2}{9c^3d}$ $\dfrac{18x^9y^5}{-6x^6y^2}$

Dividing a Binomial or Trinomial by a Monomial

To divide a binomial or trinomial by a monomial, divide each term and combine the separate quotients. Remove parentheses by dividing each term within parentheses by the monomial.

EXAMPLE 1 Divide: $\dfrac{(4y^3 + 6y)}{2y}$

STEP 1 Divide each term of the binomial by $2y$. Keep the $+$ sign that is between the terms in the binomial.

$$\frac{(4y^3 + 6y)}{2y} \rightarrow \frac{4y^3}{2y} + \frac{6y}{2y}$$

STEP 2 Divide each pair of terms.

$$\frac{4y^3}{2y} + \frac{6y}{2y} = 2y^2 + 3$$

ANSWER: $2y^2 + 3$

EXAMPLE 2 Divide: $\dfrac{(9a^2b^4 - 3ab^3 + 6b^2)}{3b^2}$

STEP 1 Divide each term of the trinomial by $3b^2$. Keep the signs that are between the terms in the trinomial.

$$\frac{(9a^2b^4 - 3ab^3 + 6b^2)}{3b^2}$$

STEP 2 Divide each pair of terms.

$$\frac{9a^2b^4}{3b^2} - \frac{3ab^3}{3b^2} + \frac{6b^2}{3b^2}$$

$$3a^2b^2 - ab + 2$$

ANSWER: $3a^2b^2 - ab + 2$

Divide. The first problem in each group is done as an example.

1. $\dfrac{(6x^4 + 4x^3)}{2x} \rightarrow 3x^3 + 2x^2$

   $\dfrac{(12n^5 + 6n^4)}{3n^2}$

   $\dfrac{(15y^7 - 9y^5)}{3y^3}$

2. $\dfrac{(-18n^5 + 6n^4)}{3n} \rightarrow -6n^4 + 2n^3$

   $\dfrac{(-20x^7 + 10x^5)}{-5x^3}$

   $\dfrac{(-14y^5 - 7y^3)}{7y^2}$

3. $\dfrac{(4a^2b^4 - 2a^2b)}{2a} \rightarrow 2ab^4 - ab$

   $\dfrac{(9m^3n + 6m^2n^2)}{3mn}$

   $\dfrac{(16c^2d^2 - 12c^2d)}{4cd}$

4. $\dfrac{(12a^3b^2 - 6a^2b^3 + 8a^2)}{2a^2} \rightarrow 6ab^2 - 3b^3 + 4$

   $\dfrac{(15m^3n^3 + 10m^2n - 5mn^2)}{5m}$

   $\dfrac{(21c^3d^2 - 7c^2d^2 + 14cd^2)}{7cd}$

Applying Your Skills: Polynomials

1. Jeni, Kendra, and Amy have tropical fish. Jeni has 4 more fish than Kendra. Amy has 2 fewer fish than Kendra. Kendra has n fish.

   a. Write a polynomial to represent how many fish Jeni and Amy have together.

   b. What is the value of this polynomial when $n = 12$?

2. An engineer uses the following polynomials to compare water flow through a pipe for two different pumps:

   Pump 1: $6r^2 - 3r + 22$

   Pump 2: $2r^2 - r + 6$

   Write a polynomial to show the difference in pumping speeds.

3. The Swanson Company makes a bread knife that has a per unit cost of $4.50 to manufacture. Also, there is a fixed cost of $3,500, regardless of the number of knives made.

   a. Write a polynomial that represents the cost of producing n knives.

   b. If the knives sell for $15, write a polynomial that shows the profit p made from selling n knives.

   c. What is p if $n = 1,000$?

4. A ticket agency charges p dollars per ticket for a concert. At this price, the agency can sell $(25,000 - p^2)$ tickets.

   a. Write a polynomial to represent the total money taken in at the concert if the ticket price is p dollars.

   b. How much money will be taken in if the ticket price is $50?

5. A rectangle is 4 feet longer than it is wide. The width of the rectangle is given as x.

   a. Write a polynomial to represent A, the area of the rectangle.

   b. What is A if $x = 8$ feet?

6. You are designing a chessboard. You label the width of the border as x.

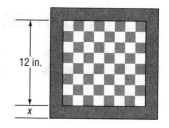

   a. Write a polynomial to represent the distance around the board.

   b. Write a polynomial to represent the area of the board.

   c. What is the area of the board when $x = 1.5$ inches?

7. A box with no top is to be made from a 20 inch by 10 inch piece of cardboard. As shown at the right, identical squares are to be cut from each corner and then the sides are to be folded up.

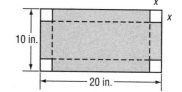

   a. Write a polynomial to represent the area of the bottom of the box.

   b. Write a polynomial to represent V, the volume of the box.

   c. What is V if $x = 1$ inch?

   d. What is V if $x = 2$ inches?

   e. What is V if $x = 3$ inches?

8. Alton Inc. took in $(500p - p^2)$ sales dollars from printer sales last month. The selling price of each printer was p.

   Write a polynomial that represents the number of printers Alton sold last month. (**Hint:** number of printers = sales dollars divided by price)

9. Jade For You sells jade jewelry. The profit from the sale of a jade ring can be represented by the polynomial $2s^2 - 12s - 74$, where s is the selling price.

   a. Write a polynomial to show the profit from the sale of 10 rings.

   b. What amount of profit is made from the sale of 10 rings if the selling price is $10?

   c. Suppose the cost of jade goes up, and the company's profit falls to half. Write a polynomial to show the reduced profit per ring if the price is kept at s.

10. A bag contains $n^3 + 4n^2 - 2n$ marbles. The marbles are divided equally among n boys.

   a. Write a polynomial to show how many marbles each boy gets.

   b. If $n = 4$, how many marbles does each boy get?

Polynomials Review

Solve each problem below. When you finish, check your answers at the back of the book. Then correct any errors.

Identify each polynomial as a monomial, binomial, or trinomial.

1. $3x + 7$ _____

2. $x^2 + 2x + 1$ _____

3. $5y^3$ _____

4. $2z^2 - 5z$ _____

Circle the pair of like terms in each group.

5. $x^2, 2x, -x^3, -5x$

6. $n^3, 3n, 2n^2, -2n^3$

7. $cd^2, -cd^2, c^2d, -c^2d^2$

8. $3ab^2, -2a^2b, a^2b^2, 3a^2b$

Add.

9. $7r + 4r$

10. $3y^2 + (-y^2)$

11. $(9x + -4) + x$

12. $(3n^2 + 5n) + (n^2 - 7n)$

Write the perimeter of each figure as a polynomial.

13.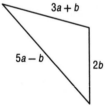

14.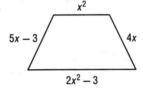

Subtract.

15. $9s - 5s$

16. $12d - (-5d)$

17. $(7a + 2b) - 4a$

18. $(12x - 3y) - (6x + 7y)$

Write each unlabeled length as a polynomial.

19.

| 3m + 2n | ? |

9m − n

20.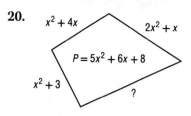

$x^2 + 4x$

$2x^2 + x$

$P = 5x^2 + 6x + 8$

$x^2 + 3$

?

Multiply.

21. $5c^4 \cdot 4c^4$

22. $-3a^2b(-2ab^3)$

23. $2x(4x^3 + 3)$

24. $2m(4m^2 + mn^2 - 3n^3)$

25. $(3a + 2)(2a - 5)$

26. $(x + 4)(x - 5)$

Write the area of each rectangle as a polynomial.

27.

5x

7x + 6

28.

3ab

$a^2 - 2b$

Divide.

29. $\dfrac{n^5}{n^3}$

30. $\dfrac{12a^7b^5}{(-3a^3b^2)}$

31. $\dfrac{(15c^4 + 6c^3)}{3c^2}$

32. $\dfrac{(14m^2n^2 - 8m^2n + 6mn^2)}{2mn}$

FACTORING

Factors

A **factor** of a number is a whole number that divides evenly into that number. The number 6 has four factors: 1, 2, 3, and 6. Every whole number greater than 1 has at least two factors: itself and the number 1.

Most whole numbers have three or more factors.
- The number 2 has only two factors: 1 and 2.
- The number 8 has four factors: 1, 2, 4, and 8.
- The number 16 has five factors: 1, 2, 4, 8, and 16.

Every whole number greater than 1 can be written as a product of two or more of its factors.

Number:	2	6	12
			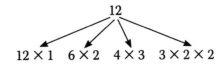
Product of Factors:	2×1	6×1 3×2	12×1 6×2 4×3 $3 \times 2 \times 2$

Write all the factors of each number. The first problem in each row is done as an example.

1. 3
   1, 3

   9

   14

2. 24
   1, 2, 3, 4, 6, 8, 12, 24

   36

   64

Write each missing factor.

3. $5 = \underline{\hspace{1cm}} \times 1$ $4 = \underline{\hspace{1cm}} \times 2$ $8 = \underline{\hspace{1cm}} \times 2$ $15 = \underline{\hspace{1cm}} \times 3$

4. $20 = \underline{\hspace{1cm}} \times 4$ $56 = \underline{\hspace{1cm}} \times 8$ $60 = \underline{\hspace{1cm}} \times 4$ $77 = \underline{\hspace{1cm}} \times 11$

Each number can be written as a product of three factors, none of which is the number 1. Write the two missing factors.

5. $8 = \underline{\hspace{1cm}} \times \underline{\hspace{1cm}} \times 2$ $18 = \underline{\hspace{1cm}} \times \underline{\hspace{1cm}} \times 2$ $20 = \underline{\hspace{1cm}} \times \underline{\hspace{1cm}} \times 2$

6. $30 = \underline{\hspace{1cm}} \times \underline{\hspace{1cm}} \times 5$ $42 = \underline{\hspace{1cm}} \times \underline{\hspace{1cm}} \times 3$ $50 = \underline{\hspace{1cm}} \times \underline{\hspace{1cm}} \times 5$

Prime Numbers and Prime Factors

A **prime number** is any whole number greater than 1 that has only two factors: itself and the number 1. Examples of prime numbers are 2, 3, 5, and 7. A **composite number** is any number greater than 1 that has more than two factors. Examples of composite numbers are 4, 6, 8, and 9.

Prime Number	Factors
2	1, 2
3	1, 3
5	1, 5
7	1, 7

Composite Number	Factors
4	1, 2, 4
6	1, 2, 3, 6
8	1, 2, 4, 8
9	1, 3, 9

The number 1 is neither a prime number nor a composite number.

Every composite number can be written as a product of **prime factors** (factors that are prime numbers). Writing a number as a product of prime numbers is called **prime-factorization** form.

Number	Find two factors.	Write each factor as a product of prime factors. (prime-factorization form)	Write each product using exponents.
8	2 and 4	$2 \times 2 \times 2$	2^3
12	2 and 6	$2 \times 2 \times 3$	$2^2 \cdot 3$
36	4 and 9	$2 \times 2 \times 3 \times 3$	$2^2 \cdot 3^2$
70	10 and 7	$2 \times 5 \times 7$	$2 \cdot 5 \cdot 7$
625	25 and 25	$5 \times 5 \times 5 \times 5$	5^4

Write all factors of each number. Circle *prime* or *composite* for each number.

1. 11

    prime or composite

2. 10

    prime or composite

3. 19

    prime or composite

4. 21

    prime or composite

5. 22

    prime or composite

6. 29

    prime or composite

Write each number in prime-factorization form or using exponents.

7. 16

8. 20

9. 32

10. 56

11. 75

12. 96

Finding the Square Root of an Algebraic Term

Sometimes you may be asked to find the square root of an algebraic term. When the term is a perfect square, the square root is easy to find.

- A perfect square can be written as the product of two identical factors. For example, $4x^2$ is a perfect square, being the product of $2x$ times $2x$. The factor $2x$ is the square root of $4x^2$.

An algebraic term is a perfect square if the coefficient is a perfect square and if each exponent is an even number.

- The coefficient of the square root is the square root of the coefficient of the term.
- Each exponent is half the value of the same exponent in the term.

To simplify the discussion on these next two pages, only positive square roots will be discussed.

> A factor of an algebraic term or expression may itself be a term or an expression.

EXAMPLE 1 Find $\sqrt{x^2}$.

ANSWER: x $(x \cdot x = x^2)$

(**Note:** $x = x^1$ and 1 is half of 2.)

EXAMPLE 3 Find $\sqrt{a^6}$.

ANSWER: a^3 $(a^3 \cdot a^3 = a^6)$

(**Note:** 3 is half of 6.)

EXAMPLE 2 Find $\sqrt{25x^6 y^4}$.

ANSWER: $5x^3y^2$ $(5x^3y^2 \cdot 5x^3y^2 = 25x^6y^4)$

(**Note:** $5 = \sqrt{25}$; 3 is half of 6; and 2 is half of 4.)

EXAMPLE 4 Find $\sqrt{16n^4}$.

ANSWER: $4n^2$ $(4n^2 \cdot 4n^2 = 16n^4)$

(**Note:** $4 = \sqrt{16}$, and 2 is half of 4.)

Find the positive square root of each algebraic term.

1. $\sqrt{y^2}$ $\qquad$ $\sqrt{n^4}$ $\qquad$ $\sqrt{y^8}$ $\qquad$ $\sqrt{r^6}$

2. $\sqrt{25x^4}$ $\qquad$ $\sqrt{16y^8}$ $\qquad$ $\sqrt{36n^6}$ $\qquad$ $\sqrt{49a^{10}}$

3. $\sqrt{4x^4y^6}$ $\qquad$ $\sqrt{25a^8b^2}$ $\qquad$ $\sqrt{81c^6d^4}$ $\qquad$ $\sqrt{64x^6y^2}$

Simplifying a Square Root

Sometimes the square root of a number or an algebraic term can only be simplified. Part of the answer remains inside the square root sign. Factoring is used to simplify the square root as much as possible. As a first step, learn this rule for square roots:

Rule: The square root of a product is equal to the product of square roots.

To see that this rule is true, consider $\sqrt{100}$ (which you know has the value 10).

STEP 1 Write 100 as 4×25 and take the square root of this product.
$$\sqrt{100} = \sqrt{4 \times 25}$$

STEP 2 Separate the square root into the product of square roots.
$$\sqrt{4} \times \sqrt{25}$$

STEP 3 Find each square root and multiply the results. Applying the rule gives the correct value 10 for $\sqrt{100}$. $2 \times 5 = \mathbf{10}$

EXAMPLE 1 Simplify: $\sqrt{90}$

STEP 1 Write 90 as $9 \cdot 10$.
$$\sqrt{90} = \sqrt{9 \cdot 10}$$

STEP 2 Separate the square root into the product of square roots.
$$\sqrt{9 \cdot 10} = \sqrt{9} \cdot \sqrt{10}$$

STEP 3 Find each square root and multiply the results.
$$\sqrt{9} \cdot \sqrt{10} = 3 \cdot \sqrt{10} = 3\sqrt{10}$$

ANSWER: $\mathbf{3\sqrt{10}}$

EXAMPLE 2 Simplify: $\sqrt{x^4 y}$

STEP 1 Write $x^4 y$ as $x^4 \cdot y$.
$$\sqrt{x^4 y} = \sqrt{x^4 \cdot y}$$

STEP 2 Separate the square root into the product of square roots.
$$\sqrt{x^4 \cdot y} = \sqrt{x^4} \cdot \sqrt{y}$$

STEP 3 Find each square root and multiply the results.
$$\sqrt{x^4} \cdot \sqrt{y} = x^2 \cdot \sqrt{y} = x^2\sqrt{y}$$

ANSWER: $\boldsymbol{x^2\sqrt{y}}$

Simplify each square root. As the first step, find a factor for each number or power that is a perfect square.

1. $\sqrt{32} = \sqrt{16} \cdot \sqrt{2}$
    $= 4\sqrt{2}$

    $\sqrt{45} =$

    $\sqrt{75} =$

2. $\sqrt{x^2 y} = \sqrt{x^2} \cdot \sqrt{y}$
    $= x\sqrt{y}$

    $\sqrt{x^6 y}$

    $\sqrt{ab^8}$

3. $\sqrt{4xy^2} = \sqrt{4} \cdot \sqrt{y^2} \cdot \sqrt{x}$
    $= 2y\sqrt{x}$

    $\sqrt{16yz^6}$

    $\sqrt{25r^4 s}$

Factoring an Algebraic Expression

If each term of an expression can be divided evenly by a number or by a variable, that number or variable is a factor of the expression. This is the reverse of earlier when you distributed multiplication through parentheses to remove the parentheses. When factoring is done, notice that parentheses are added.

Factoring Out a Number

EXAMPLE 1 Write $4x + 10$ as a product of factors.

> To factor an algebraic expression is to write the expression as a product of factors.

STEP 1 Find the greatest number that divides evenly into each term ($4x$ and 10). The number is 2.

STEP 2 Write 2 on the outside of a set of parentheses. The result of dividing each term by 2 goes inside the parentheses.

ANSWER: $4x + 10 \longrightarrow 2(2x + 5)$

Factoring Out a Variable

EXAMPLE 2 Write $5x^2 + 3x$ as a product of factors.

> The expression $5x^2 + 3x$ contains two terms. Because there is no number that divides evenly into both coefficients, no number can be factored out.

STEP 1 Each term can be divided evenly by x. So, x can be factored out.

STEP 2 Write x on the outside of a set of parentheses. For the inside terms, divide $5x^2$ by x $\left(\frac{5x^2}{x} = 5x\right)$ and divide $3x$ by x $\left(\frac{3x}{x} = 3\right)$.

ANSWER: $5x^2 + 3x \longrightarrow x(5x + 3)$

Factor a number out of each algebraic expression.

1. $2x + 6 \longrightarrow 2(x + 3)$ $3x + 12$ $5y - 20$
   (factor out a 2)

2. $3x^2 + 6x + 3 \longrightarrow 3(x^2 + 2x + 1)$ $4x^2 - 8x + 12$ $7y^2 - 14y - 35$
   (factor out a 3)

Factor a variable out of each algebraic expression.

3. $2x^2 + 5x \longrightarrow x(2x + 5)$ $3x^2 - 7x$ $2z^2 + 11z$
   (factor out an x)

4. $y^3 - 2y^2 + 4y \longrightarrow y(y^2 - 2y + 4)$ $z^3 + 6z^2 - 4z$ $x^3 - 3x^2 + 2x$
   (factor out a y)

Factoring Out a Number *and* a Variable

EXAMPLE 3 Write $4x^2 - 12x$ as a product of factors.

STEP 1 Find the greatest number that divides evenly into each term. This number is 4.

STEP 2 Notice that you can factor an x out of each term. Factor out both 4 and x. To do this, factor out the product $4x$.

ANSWER: $4x^2 - 12x \longrightarrow 4x(x - 3)$

> In many algebraic expressions, both numbers and variables can be factored out.

EXAMPLE 4 Write $3x^3y + 15x^2y$ as a product of factors.

STEP 1 Find the greatest number that divides evenly into each term. This number is 3.

STEP 2 Notice that you can also factor the product x^2y out of each term. Factor out $3x^2y$.

ANSWER: $3x^3y + 15x^2y \longrightarrow 3x^2y(x + 5)$

> A product of variables may be a factor in an algebraic expression.

Factor each algebraic expression as completely as possible.

5. $2w^2 + 8 \longrightarrow 2(w^2 + 4)$
   (factor out 2)

   $9x^2 + 3$

   $12y^2 - 8$

6. $5x^2 + 10x \longrightarrow 5x(x + 2)$
   (factor out $5x$)

   $6x^2 + 24x$

   $7z^2 - 21z$

7. $12z^2 - 18z \longrightarrow 6z(2z - 3)$
   (factor out $6z$)

   $4y^2 + 6y$

   $10z^2 - 15z$

8. $3y^3 + 15y^2 \longrightarrow 3y^2(y + 5)$
   (factor out $3y^2$)

   $5x^3 - 25x^2$

   $4s^3 + 16s^2$

9. $8x^4 + 12x^2 \longrightarrow 4x^2(2x^2 + 3)$
   (factor out $4x^2$)

   $9y^4 + 6y^2$

   $15z^4 + 20z^2$

10. $2ab^2 + 4a^2b \longrightarrow 2ab(b + 2a)$
    (factor out $2ab$)

    $6x^2y + 3xy$

    $4r^2s + 6rs^2$

Factoring and Dividing

Factoring is often used to simplify a quotient of algebraic expressions. If a single term or an expression of 2 or more terms is a factor in both top and bottom expressions, that term or expression divides out.

EXAMPLE 1 Divide: $\dfrac{6n^2 + 2}{2}$

STEP 1 Factor 2 from $6n^2 + 2$.

$6n^2 + 2 \longrightarrow 2(3n^2 + 1)$

STEP 2 Divide by 2.

$\dfrac{6n^2 + 2}{2} \longrightarrow \dfrac{\cancel{2}(3n^2 + 1)}{\cancel{2}}$

ANSWER: $3n^2 + 1$

EXAMPLE 2 Divide: $\dfrac{3x^2 + 6x}{3x}$

STEP 1 Factor $3x$ from $3x^2 + 6x$.

$3x^2 + 6x \longrightarrow 3x(x + 2)$

STEP 2 Divide by $3x$.

$\dfrac{3x^2 + 6x}{3x} \longrightarrow \dfrac{\cancel{3x}(x + 2)}{\cancel{3x}}$

ANSWER: $x + 2$

EXAMPLE 3 Divide: $\dfrac{8x^2y^2 - 4xy}{4xy}$

STEP 1 Factor $4xy$ from $8x^2y^2 - 4xy$.

$8x^2y^2 - 4xy \longrightarrow 4xy(2xy - 1)$

STEP 2 Divide by $4xy$.

$\dfrac{8x^2y^2 - 4xy}{4xy} \longrightarrow \dfrac{\cancel{4xy}(2xy - 1)}{\cancel{4xy}}$

ANSWER: $2xy - 1$

EXAMPLE 4 Divide: $\dfrac{2a^2 + 2ab}{a + b}$

STEP 1 Factor $2a$ from $2a^2 + 2ab$.

$2a^2 + 2ab \longrightarrow 2a(a + b)$

STEP 2 Divide by $a + b$.

$\dfrac{2a^2 + 2ab}{a + b} \longrightarrow \dfrac{2a\cancel{(a + b)}}{\cancel{a + b}}$

ANSWER: $2a$

Simplify each quotient by dividing out the common factor. The first problem in each row is started for you.

1. $\dfrac{4x^2 + 8}{4} \longrightarrow \dfrac{4(\underline{\hspace{1cm}})}{4} \longrightarrow$

   $\dfrac{8r^2 + 6}{2}$

2. $\dfrac{4y^2 + 2y}{2y} \longrightarrow \dfrac{2y(\underline{\hspace{1cm}})}{2y} \longrightarrow$

   $\dfrac{2n^2 + 6n}{2n}$

3. $\dfrac{5x^2y - 10xy}{5xy} \longrightarrow \dfrac{5xy(\underline{\hspace{1cm}})}{5xy} \longrightarrow$

   $\dfrac{6ab - 3a^2b^2}{3ab}$

4. $\dfrac{3x + 3y}{x + y} \rightarrow \dfrac{3()}{x + y} \rightarrow$ $\dfrac{4c - 4d}{c - d}$

5. $\dfrac{x^2 + xy}{x + y} \rightarrow \dfrac{x()}{x + y} \rightarrow$ $\dfrac{mn + n^2}{m + n}$

6. $\dfrac{3r^2 - 3rs}{r - s} \rightarrow \dfrac{3r()}{r - s} \rightarrow$ $\dfrac{7uv + 7v^2}{u + v}$

Write each missing dimension as a simplified algebraic expression.

7. length = _____

xy | $A = 4x^2y$

length = ?

$$\text{length} = \frac{\text{area}}{\text{width}}$$

8. width = _____

width = ? | $A = 6a^2b + 3a^2b^2$

$3a^2b$

$$\text{width} = \frac{\text{area}}{\text{length}}$$

9. height = _____

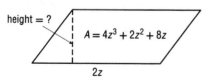

height = ? $A = 4z^3 + 2z^2 + 8z$

$2z$

$$\text{height} = \frac{\text{area}}{\text{base}}$$

10. height = _____

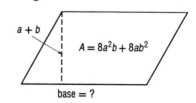

$a + b$ $A = 8a^2b + 8ab^2$

base = ?

$$\text{height} = \frac{\text{area}}{\text{base}}$$

Factoring a Difference of Squares

When you multiply the sum of two variables by the difference of the same two variables, the product is the difference of the squares of the variables.

$(a + b)(a - b) = a^2 - b^2$ (Recall that $(a + b)(a - b) = a^2 + ab - ab - b^2 = a^2 - b^2$)

Knowing this enables you to write the rule for factoring a difference of squares.

Rule for factoring a difference of squares:

A difference of squares factors into two binomials: the sum of the variables times the difference of the variables.

This rule applies to numbers as well as variables.

EXAMPLE 1 Factor $x^2 - 4$.

STEP 1   Find the square roots of x^2 and 4.

$\sqrt{x^2} = x$ and $\sqrt{4} = 2$

STEP 2   Write the difference of squares as a product of binomial factors.

$x^2 - 4 \longrightarrow (x + 2)(x - 2)$

ANSWER: $x^2 - 4 \longrightarrow (x + 2)(x - 2)$

EXAMPLE 2 Factor $16 - r^2$.

STEP 1   Find the square roots of 16 and r^2.

$\sqrt{16} = 4$ and $\sqrt{r^2} = r$

STEP 2   Write the difference of squares as a product of binomial factors.

$16 - r^2 \longrightarrow (4 + r)(4 - r)$

ANSWER: $16 - r^2 \longrightarrow (4 + r)(4 - r)$

Factor each difference of squares.

1. $25 - s^2$ $36 - t^2$

2. $64 - n^2$ $16 - x^2$

3. $r^2 - 49$ $y^2 - 81$

Completely factor each expression. The first problem in each row is done as an example.

4. $x^3 - 25x \rightarrow x(x^2 - 25) \rightarrow x(x + 5)(x - 5)$

> For problems 4 and 5, a term is factored out and then the difference of squares is factored.

$y^3 - 36y$

$n^3 - 16n$

5. $4s^3 - 36s \rightarrow 4s(s^2 - 9) \rightarrow 4s(s + 3)(s - 3)$

$8x^3 - 32x$

$2y^3 - 32y$

Simplify each quotient by factoring. The first problem is done as an example.

6. $\dfrac{(d + 9)(d - 9)}{d - 9} \rightarrow \dfrac{(d + 9)\,\cancel{(d - 9)}}{\cancel{d - 9}} \rightarrow d + 9$ $\qquad \dfrac{t^2 - 64}{t + 8}$

7. $\dfrac{25 - r^2}{5 + r}$ $\qquad\qquad\qquad\qquad\qquad \dfrac{4n^3 - 36n}{n - 3}$

Write each missing dimension as a simplified algebraic expression. As a first step, completely factor the expression for each area.

8. length = _____

$(7 - x)$ | $A = 49 - x^2$ | length = ?

9. width = _____

width = ? | $A = a^2 - 64$ | $a + 8$

10. height = _____

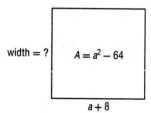
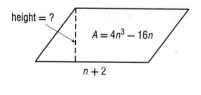

height = ? $\quad A = 4n^3 - 16n$ $\quad n + 2$

length = $\dfrac{\text{area}}{\text{width}}$ $\qquad\qquad$ width = $\dfrac{\text{area}}{\text{length}}$ $\qquad\qquad$ height = $\dfrac{\text{area}}{\text{base}}$

Factoring Review

Solve the problems below. When you finish, check your answers at the back of the book. Then correct any errors.

Write all the factors of each number.

1. 9

2. 7

3. 24

Each number can be written as a product of three factors, none of which is the number 1. Write the two missing factors.

4. $24 = \underline{\quad\quad} \times \underline{\quad\quad} \times 3$

5. $30 = \underline{\quad\quad} \times \underline{\quad\quad} \times 5$

6. $42 = \underline{\quad\quad} \times \underline{\quad\quad} \times 2$

Write each number in prime-factorization form.

7. 14

8. 16

9. 50

Find the positive square root of each term.

10. $\sqrt{16x^4}$

11. $\sqrt{36a^2b^2}$

12. $\sqrt{25c^4d^6}$

Simplify each square root.

13. $\sqrt{18}$

14. $\sqrt{25x^2y}$

15. $\sqrt{49cd^6}$

Factor a number out of each expression.

16. $4x + 4$

17. $3y - 15$

18. $6y + 24$

Factor a variable out of each expression.

19. $3x^2 + 7x$

20. $2n^3 - 9n$

21. $5y^4 + 2y^2$

Factor each expression as completely as possible.

22. $3x^2 + 12x$

23. $5n^3 - 15n^2$

24. $8x^2y + 2xy$

Simplify each quotient by dividing out the common factor.

25. $\dfrac{4x + 4}{4}$

26. $\dfrac{3n^2 + 6}{3}$

27. $\dfrac{6m - 3m^2n}{3m}$

28. $\dfrac{4x + 4y}{x + y}$

29. $\dfrac{xy + y^2}{x + y}$

30. $\dfrac{3ab - 3b^2}{a - b}$

Factor each difference of squares.

31. $36 - n^2$

32. $x^2 - 25$

Simplify each quotient by factoring.

33. $\dfrac{x^2 - 49}{x + 7}$

34. $\dfrac{3n^2 - 27}{n - 3}$

Write each missing dimension as a simplified algebraic expression.

35. length = _____

4ab | $A = 12a^2b^2 + 4ab^2$

length = ?

36. width = _____

width = ? | $A = x^2 - 36$

$x + 6$

Algebra Posttest A

This posttest gives you a chance to check the skills you have learned in *Algebra*. Take your time and work each problem carefully. When you finish, check your answers and review any topics on which you need more work.

Write the addition shown by the number arrows.

1. _____

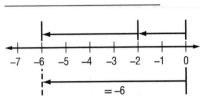

2. _____

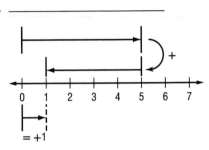

Subtract.

3. $8 - (-7) =$

4. $-4 - (-7) =$

Multiply or divide as indicated.

5. $(-8)(4) =$

6. $\dfrac{-5}{-9} =$

Find the value of each power.

7. $3^4 =$

8. $4^{-3} =$

Simplify each product or quotient by writing as a single power.

9. $(7^3)(7^2)(7) =$

10. $\dfrac{10^8}{10^4} =$

Find each square root below.

11. $\sqrt{81} =$

12. $\sqrt{\dfrac{36}{25}} =$

13. Write the expression "$\dfrac{1}{4}$ times the quantity x minus 6" as an algebraic expression.

Evaluate each expression.

14. $5n^2 - 3n$ when $n = 2$

15. $ab\left(\dfrac{a+4}{b-4}\right)$ when $a = -2, b = 6$

16. Use the formula $°C = \dfrac{5}{9}(°F - 32)$ to find the Celsius temperature (°C) when the Fahrenheit temperature (°F) is 77°F.

Circle the equation that is *not* represented in the drawing.

17.

a. $n - 4 = 10$

b. $10 - n = 4$

c. $n + 4 = 10$

d. $10 - 4 = n$

For problems 18–22, solve each equation. Show all steps.

18. a. $n + 12 = 6$

b. $x - 7 = 30$

19. a. $4p = 36$

b. $\frac{r}{6} = 11$

20. a. $4x - 7 = 29$

b. $\frac{a}{3} + 4 = 9$

21. a. $6n + 2n = 72$

b. $5x - 8 = x + 16$

22. a. $3(s - 2) = 33$

b. $5(y - 2) = 2(y + 13)$

23. A field is in the shape of a rectangle. The ratio of the length of the field to its width is 3 to 2. If the perimeter of the field is 190 yards, what are the dimensions of the field?

24. In a survey of 360 high school students, 3 out of every 5 said they ride the school bus at least four times each week. Use a proportion to find the number of students who do *not* ride the school bus at least four times each week.

25. Find the two correct values of x in the equation $4x^2 = y$ when $y = 144$.

26. Graph the line containing the points $(-3, 0)$, $(0, 1)$, and $(3, 2)$.

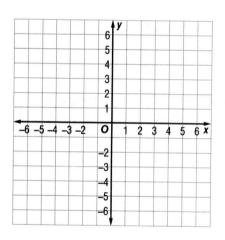

27. What is the slope of line R?

$$\text{slope} = \frac{\text{change in } y}{\text{change in } x}$$

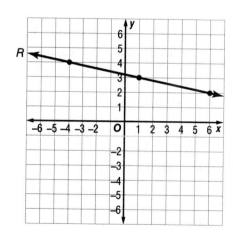

28. Graph the equation $y = 3x - 4$. As a first step, complete the Table of Values.

Table of Values	
x	**y**
0	
2	
3	

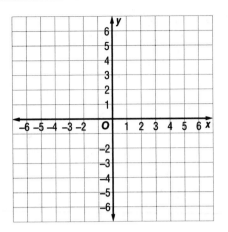

Write the inequality graphed on the number line.

29.

Values of n

Solve the following inequalities.

30. $x - 8 \geq -2$

31. $4r + 7 < 35$

Graph the inequality $-4 \leq s < 5$.

32.

Values of s

33. Circle the pair of like terms.

$2y^3 \qquad 2x^3 \qquad 3y^2 \qquad -3x^2 \qquad -3x^3$

Add or subtract as indicated.

34. $(-4n^2 + 7) + (5n^2 - 9)$

35. $(5x^2 - 6x) - (x^2 + 3x)$

Multiply or divide as indicated.

36. $4d(2d^2 - d + 1)$

37. $\dfrac{(21n^4 + 15n^2)}{3n}$

Simplify each square root.

38. $\sqrt{72} =$

39. $\sqrt{49a^6b}$

Factor each expression as completely as possible.

40. $9x^2 + 12$

41. $20n^3 + 15n^2$

Simplify each quotient by dividing out the common factor.

42. $\dfrac{8x^2 + 4}{4}$

43. $\dfrac{9a - 9b}{a - b}$

ALGEBRA POSTTEST A CHART

Circle the number of any problem that you miss. A passing score is 37 correct answers. If you passed the test, go on to Using Number Power. If you did not pass the test, review the chapters in this book or refer to the practice pages listed for each skill area.

Problem Numbers	Skill Area	Practice Pages
1, 2, 3, 4, 5, 6	signed numbers	10–23
7, 8, 9, 10, 11, 12	powers and roots	26–39
13, 14, 15, 16	algebraic expressions	42–53
17, 18, 19	one-step equations	56–67
20, 21, 22	multistep equations	72–87
23, 24, 25	ratios, proportions, and quadratics	92–105
26, 27, 28	graphing equations	108–121
29, 30, 31, 32	inequalities	124–133
33, 34, 35, 36, 37	polynomials	136–149
38, 39, 40, 41, 42, 43	factoring	152–161

USING
NUMBER
POWER

Windchill

When the wind blows, your body feels colder than the actual temperature. For example, a wind speed of 25 miles per hour causes a 10°F temperature to feel as cold as a temperature of −11°F in which no wind is blowing. **Windchill** is the name given to the temperature that your body actually feels when the wind is blowing.

The table below shows thermometer temperatures and the windchill temperature for various wind speeds. The example given above is highlighted in the table.

		Thermometer Temperature (°F)										
		−25°	−20°	−15°	−10°	−5°	0°	5°	10°	15°	20°	25°
		Windchill Temperature (°F)										
Wind Speed (mph)	5	−40°	−34°	−28°	−22°	−16°	−11°	−5°	1°	7°	13°	19°
	10	−47°	−41°	−35°	−28°	−22°	−16°	−10°	−4°	3°	9°	15°
	15	−51°	−45°	−39°	−32°	−26°	−19°	−13°	−7°	0°	6°	13°
	20	−55°	−48°	−42°	−35°	−29°	−22°	−15°	−9°	−2°	4°	11°
	25	−58°	−51°	−44°	−37°	−31°	−24°	−17°	−11°	−4°	3°	9°
	30	−60°	−53°	−46°	−39°	−33°	−26°	−19°	−12°	−5°	1°	8°
	35	−62°	−55°	−48°	−41°	−34°	−27°	−21°	−14°	−7°	0°	7°
	40	−64°	−57°	−50°	−43°	−36°	−29°	−22°	−15°	−8°	−1°	6°

EXAMPLE When the wind speed is 35 mph, how many degrees below the thermometer temperature of 25°F is the windchill temperature?

STEP 1 On the table, read down the column beneath the temperature 25°F. Read across from the row of wind speed 35 mph until it meets the 25° temperature column. This temperature is 7°F.

STEP 2 Subtract 7 from 25.

25 − 7 = **18**

ANSWER: The windchill temperature is **18°** below the thermometer temperature.

Use the table on page 172 to find the windchill temperature for each thermometer temperature.

1. thermometer temperature = −20°F
   wind speed = 30 mph

   windchill temperature =

2. thermometer temperature = −5°F
   wind speed = 10 mph

   windchill temperature =

3. thermometer temperature = 0°F
   wind speed = 40 mph

   windchill temperature =

4. thermometer temperature = 20°F
   wind speed = 15 mph

   windchill temperature =

Use the table on page 172 to find the difference between the thermometer temperature and the windchill temperature (thermometer temperature minus windchill temperature). As a first step, find the windchill temperature.

5. thermometer temperature = −15°F
   wind speed = 20 mph

   difference between thermometer
   and windchill temperatures =

6. thermometer temperature = 15°F
   wind speed = 30 mph

   difference between thermometer
   and windchill temperatures =

Use the table on page 172 to solve each problem.

7. A weather reporter in Chicago says that the windchill temperature is −7°F. At the same time, Kristen reads an outdoor thermometer which gives a temperature of 15°F. At this moment, what is the wind speed?

8. While standing in a 10 mph wind, Sean hears a report that the windchill temperature is −22°F. At this moment, what is the actual thermometer temperature?

Distance Formula

The distance formula, $d = rt$, tells how to find the distance traveled d when you know the rate r and the time t of travel.

EXAMPLE 1 How far can Joan drive in 7 hours if she averages 50 miles per hour?

  STEP 1 Identify r and t.

  $r = 50$ miles per hour and $t = 7$ hours

  STEP 2 Substitute the r and t values into the distance formula.

  $d = rt = 50 \times 7 = $ **350 miles**

ANSWER: 350 miles

If the distance and time are known, you can find the rate (speed) by solving the distance formula for r to get $r = \frac{d}{t}$.

EXAMPLE 2 On part of her trip, Joan drove 405 miles in 9 hours. What was her average speed?

  STEP 1 Identify d and t.

  $d = 405$ miles and $t = 9$ hours

  STEP 2 Substitute the d and t values into the new formula.

  $r = \frac{d}{t} = \frac{405}{9} = $ **45 miles per hour**

ANSWER: 45 miles per hour

You can also use the distance formula to find the time when the distance and rate (speed) are known. Solve the distance formula for t to get $t = \frac{d}{r}$.

EXAMPLE 3 Joan has 280 miles left to drive. If she averages 40 miles per hour, how long will it take her to drive this distance?

  STEP 1 Identify d and r.

  $d = 280$ miles and $r = 40$ miles per hour

  STEP 2 Substitute the d and r values into the new formula.

  $t = \frac{d}{r} = \frac{280}{40} = $ **7 hours**

ANSWER: 7 hours

For each problem, decide whether you are finding the distance *d*, rate *r*, or time *t*. Then use the appropriate formula to solve.

1. What distance can Nick drive in 12 hours if he averages 50 miles per hour?

2. On the second day of his trip, Nick drove 440 miles in 8 hours. Find his average speed during the second day.

3. If Nick can average 50 miles per hour on the third day, how long will it take him to drive 350 miles?

4. Jennifer drove from Chicago to New York. On the first day of her trip, she averaged 48 miles per hour for 6 hours. How far did she drive the first day?

5. On the second day of her trip, Jennifer drove 364 miles in 7 hours. Find her average speed during the second day.

6. On the third day, Jennifer averaged 54 miles per hour as she drove 324 miles. How many hours did Jennifer drive on the third day?

Simple Interest Formula

Interest is money that is earned (or paid) for the use of someone else's money.
- If you deposit money in a savings account, interest is the money the bank pays you for using your money.
- If you borrow money, interest is the money you pay for using the lender's money.

Simple interest is interest on a principal amount—money deposited or borrowed. To compute simple interest, use the **simple interest formula.** The formula is $i = prt$, where

i = simple interest, in dollars
p = principal, money deposited or borrowed, in dollars
r = percent rate, as a fraction or decimal
t = time, in years

EXAMPLE 1 What is the simple interest earned on $300 deposited for 2 years in a savings account which pays 5% simple interest?

STEP 1 Identify p, r, and t.

$$p = 300 \qquad r = 5\% = \frac{5}{100} \qquad t = 2$$

The rate, 5%, is expressed as the fraction $\frac{5}{100}$ to simplify multiplication in Step 2.

STEP 2 Substitute and multiply.

$$i = prt = \overset{3}{\cancel{300}} \times \frac{5}{\underset{1}{\cancel{100}}} \times 2 = \textbf{\$30}$$

ANSWER: **$30**

EXAMPLE 2 At the end of 3 years, what is the total amount owed on a $429 loan borrowed at 9.5% simple interest per year?

STEP 1 Identify p, r, and t.

$$p = 429 \qquad r = 9.5\% = 0.095 \qquad t = 3$$

Sometimes it is easier to multiply with decimals than with fractions.

STEP 2 Substitute and multiply.

$pr = 429 \times 0.095 = 40.755$

$prt = 40.755 \times 3 = \$122.265 \approx \$122.27$ (nearest cent)

STEP 3 principal + interest = total amount owed

$\$429.00 + 122.27 = \textbf{\$551.27}$

ANSWER: **$551.27**

Use the simple interest formula to solve each problem.

1. What is the interest earned on $200 deposited for 3 years in a savings account that pays 5% simple interest?

2. Jane deposited $650 at a $5\frac{1}{2}$% simple interest rate. What will be the total amount in her account after 2 years?

3. At a 10% interest rate, how much simple interest will Bill pay on a loan of $475 borrowed for 2 years?

4. What is the total amount owed on a loan if the principal is $1,000, the simple interest rate is 11%, and the time is 3 years?

5. How much simple interest will John pay for a loan of $575 at 12.5% if he repays the bank at the end of 1 year?

6. What is the total amount in a savings account if the amount deposited was $375, the simple interest rate was 5.25%, and the time was 4 years?

Simple Interest Formula: Parts of a Year

Interest is paid on the basis of a yearly percent rate. However, not all deposits or loans are made for whole years. To use the simple interest formula for a time period that is not a whole year, write the time as a fraction of a year.

EXAMPLE 1   How much simple interest is earned on a $600 deposit at an annual interest rate of 7% for 8 months?

STEP 1   Identify p, r, and t.

$$p = 600 \qquad r = \frac{7}{100} \qquad t = \frac{8}{12} = \frac{2}{3}$$

(**Note:** 8 months $= \frac{8}{12}$ year $= \frac{2}{3}$ year)

STEP 2   Substitute and multiply.

$$i = prt = \overset{6}{\cancel{600}}\left(\frac{7}{\underset{1}{\cancel{100}}}\right)\left(\frac{2}{3}\right) = 6(7)\left(\frac{2}{3}\right) = \textbf{\$28}$$

ANSWER: \$28

EXAMPLE 2   What is the interest paid on a loan of $800 borrowed for 2 years and 3 months at 9% simple interest?

STEP 1   Identify p, r, and t.

$$p = 800 \qquad r = \frac{9}{100} \qquad t = 2\frac{3}{12} = 2\frac{1}{4} = \frac{9}{4}$$

(**Note:** Time is written as $\frac{9}{4}$ to simplify multiplication in Step 2.)

STEP 2   Substitute and multiply.

$$i = prt = \overset{8}{\cancel{800}}\left(\frac{9}{\underset{1}{\cancel{100}}}\right)\left(\frac{9}{4}\right) = 8(9)\left(\frac{9}{4}\right) = \textbf{\$162}$$

ANSWER: \$162

EXAMPLE 3   What is the interest earned on $200 deposited for 16 months in a savings account paying 6% simple interest?

STEP 1   Identify p, r, and t.

$$p = 200 \qquad r = \frac{6}{100} \qquad t = \frac{16}{12} = \frac{4}{3}$$

STEP 2   Substitute and multiply.

$$i = prt = \overset{2}{\cancel{200}}\left(\frac{6}{\underset{1}{\cancel{100}}}\right)\left(\frac{4}{3}\right) = 2(6)\left(\frac{4}{3}\right) = \textbf{\$16}$$

ANSWER: \$16

Use the simple interest formula to solve each problem.

1. What is the interest earned on $900 deposited for 1 year 6 months in a savings account that pays 5% simple interest?

2. How much money will Tracy repay to the bank for a loan of $1,000 that she borrows for 2 years and 8 months at a 12% simple interest rate?

3. Eva deposited $850 in a savings account. How much interest will her money earn after 20 months if the simple interest rate is 6%?

4. How much money will Lam repay to the bank if he borrows $1,000 for 9 months at a simple interest rate of 16%?

5. Top Dollar Visa is charging Keisha 18% simple interest on her $750 unpaid bill. How much interest will Keisha pay if she pays the entire bill after 10 months?

6. Anne plans to take out a $7,000 loan at Evergreen Bank to buy a used car. The bank agrees to let Anne pay the principal plus simple interest at the end of 18 months. What total amount will Anne owe?

Evergreen Bank Simple Interest Loans	
New car	9.9%
Used car	10.5%
Boat	12.8%

Work Problems

A common type of problem (often called a **work problem**) involves finding the rate at which two or more people working together can do a job. If you know the rate at which each person works alone, you can find the rate at which they work together.

EXAMPLE Jim can do a yard job alone in 2 hours. Miah can do the same job alone in 3 hours. How long will it take both of them working together to finish the job?

STEP 1 Express the fraction of the job each can do in 1 hour.

$\frac{1}{2}$ = fraction of the job Jim can do in 1 hour. (If Jim can do the whole job in 2 hours, he can do $\frac{1}{2}$ of the job in 1 hour.)

$\frac{1}{3}$ = fraction of the job Miah can do in 1 hour

STEP 2 Express the fraction of the job each can do in x hours, where x equals the time it takes to complete the job together.

$\frac{1}{2}x$ = fraction of the job Jim can do in x hours

$\frac{1}{3}x$ = fraction of the job Miah can do in x hours

STEP 3 Set the sum of the two fractions equal to 1 and solve the resulting equation for x.

(**Note:** Step 3 says that the sum of the parts of the job, the fractions, is equal to the whole job, represented by the number 1.)

a. Add the x terms. $\frac{1}{2}x + \frac{1}{3}x = 1$

$\frac{1}{2} + \frac{1}{3} = \frac{3}{6} + \frac{2}{6} = \frac{5}{6}$ $\frac{5}{6}x = 1$

b. Multiply $\frac{5}{6}$ by its reciprocal of $\frac{6}{5}$ so that x stands alone.

$\frac{6}{5}\left(\frac{5}{6}\right)x = \frac{6}{5}(1)$

$x = \frac{6}{5} = \mathbf{1\frac{1}{5}}$

ANSWER: $1\frac{1}{5}$ hr = 1 hr 12 min

The steps used to solve the example can be summarized as follows.

STEP 1 Find the fractional part of the job each person can do in 1 hour.

STEP 2 Multiply each fraction from Step 1 by x.

STEP 3 Set the sum of the fractions equal to 1 and solve for x.

Solve.

1. Kami can wash her apartment windows in 1 hour if she works alone. Jeff, working alone, also takes just 1 hour to do the same windows. If they work together, how long will it take them to wash the windows?

2. By herself, Mariah can trim the fruit trees in 2 hours. Paula can do the same job by herself in 4 hours. Working together, how long will it take Mariah and Paula to trim the trees?

3. Working alone, Carrie can paint a room in 4 hours. Amy, working alone, can paint the same room in 3 hours. How long will it take both of them working together to paint the room?

4. Georgi can wax a large truck in 5 hours when he is working alone. Jerry can do the same job alone in 4 hours. How long will it take Georgi and Jerry to wax the truck if they work together?

5. Using the larger tractor, Warren can plow a field in 3 hours. Using the smaller tractor, Eva can plow the same field in 5 hours. How long will it take them to plow the field working together?

Mixture Problems

Do you ever wonder how the prices are determined for different mixtures of nuts or candies? For example, you know that a mixture of cashews and peanuts costs more than peanuts alone but costs less than cashews alone. How is the exact price determined? With algebra, you can answer this question.

EXAMPLE How many pounds of a $5-per-pound mixture of cashews should be mixed with 3 pounds of a $2-per-pound mixture of peanuts to obtain a new mixture with a value of $4 per pound?

STEP 1 Let x = the amount of $5-per-pound mixture of nuts needed. Put the information you know in a table as follows:

a. Row A is the cost per pound of each mixture.

b. Row B is the number of pounds of each mixture.

c. Row C is the dollar value of each mixture: ($).
($) = cost per pound times number of pounds

		Cashews		Peanuts		Mixture
Row A	cost per pound:	$5		$2		$4
× Row B	× number of pounds:	× x		× 3		$x + 3$
Row C	**value ($)**	**5x**	+	**6**	=	**4(x + 3)**
(Row C = Row A × Row B)						

Row C gives the equation to solve: $5x + 6 = 4(x + 3)$.
The value of the $5-per-pound mixture ($5x$) plus the value of the $2-per-pound mixture ($2 × 3 = $6) is equal to the value of the $4-per-pound mixture $4(x + 3)$.

In words: value of $5 mixture + value of $2 mixture = value of $4 mixture

In symbols: $5x + 6 = 4(x + 3)$

STEP 2 Solve the equation for x. $5x + 6 = 4(x + 3)$

a. Multiply by 4 to remove the parentheses. $5x + 6 = 4x + 12$

b. Subtract 6 from both sides. $5x = 4x + 12 - 6$
     $5x = 4x + 6$

c. Subtract $4x$ from both sides. $5x - 4x = 6$
     $x = 6$

ANSWER: **6 pounds** of the $5-per-pound mixture of cashews should be used.

Solve.

1. How many pounds of a $6-per-pound mixture of nuts should be mixed with 5 pounds of a $3-per-pound mixture of candy to obtain a mixture of candy and nuts costing $5 per pound?

2. How many pounds of a dried fruit mix costing $3.00 per pound should be mixed with 8 pounds of a nut mixture costing $4.50 per pound to give a fruit-nut mixture costing $4.00 per pound?

3. How many pounds of chocolates costing $2.00 per pound should be mixed with 3 pounds of chocolates costing $1.60 per pound to obtain a mixture of chocolates costing $1.70 per pound?

Temperature: Finding Fahrenheit from Celsius

The temperature formula introduced on page 117 can be rewritten without parentheses so that °F appears on the left side of the equal sign.

$$°F = \frac{9}{5}°C + 32°$$

In this form, you can use the formula to find an equivalent Fahrenheit temperature for a given Celsius temperature.

EXAMPLE Find the Fahrenheit temperature when the Celsius temperature is 40°.

$$°F = \frac{9}{5}°C + 32$$

STEP 1 Substitute 40° for °C.

$$°F = \frac{9}{5}(40) + 32$$

STEP 2 Multiply 40 by $\frac{9}{5}$.

$$\frac{^8\cancel{40}}{1} \times \frac{9}{\cancel{5}_1} = 72$$

STEP 3 Add.

$$°F = 72 + 32 = \mathbf{104°}$$

ANSWER: **104°F**

For problems 1–5, use the temperature formula $°F = \frac{9}{5}°C + 32°$.

1. Water freezes at 0° Celsius. What is the Fahrenheit temperature at which water freezes?

2. Mary likes to be outdoors when the temperature reaches 25°C. What would be the equivalent Fahrenheit temperature?

3. What is the oven temperature in °F when meat is broiling at 180°C?

4. When Raul had the flu, his temperature was 38°C. What was his temperature in °F?

5. A comfortable room temperature is 20°C. What is the equivalent temperature in °F?

Use the temperature formula to complete the Table of Values.

- **Fill in the Points to Graph with the coordinates of the three points identified in the completed Table of Values.**
- **Plot the three points on the graph.**
- **Connect the points with a straight line and extend the line to the edge of the graph.**

6.

Table of Values

°C	°F
0	
10	
30	

Points to Graph

(,)

(,)

(,)

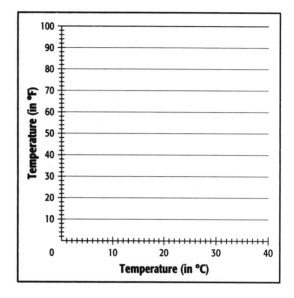

7. Use the graph to find the approximate temperature, in °F, when the temperature is 28°C.

8. By reading the graph, find the approximate temperature, in °C, when the temperature is 100°F.

9. What is the slope of the temperature line?

Using Algebra in Geometry

Algebra is often used to solve problems in geometry. In the examples below, unknown angles are represented by variables, and equations are used to find the measure of each angle.

From geometry, you know that an angle ($\angle$) is measured in units called **degrees** (°). A right angle ($\llcorner$), formed by perpendicular lines, measures 90°. An angle can be labeled with three letters, the second letter being the **vertex,** the point where the two sides meet.

EXAMPLE 1 $\angle ABC$ is a right angle (90°) that is divided into three smaller angles.

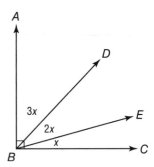

$\angle DBE = 2$ times $\angle EBC$, or $2\angle EBC$ and

$\angle ABD = 3$ times $\angle EBC$, or $3\angle EBC$.

What is the measure of each of these smaller angles?
(**Note:** m$\angle$ is read as 'the measure of angle'.)

STEP 1 Let m$\angle EBC = x$, m$\angle DBE = 2x$, and m$\angle ABD = 3x$.

STEP 2 Set the sum of the angles equal to 90° and solve the equation.

$$x + 2x + 3x = 90$$
$$6x = 90$$
$$x = \frac{90}{6} = \textbf{15}$$
$$2x = 2(15) = \textbf{30}$$
$$3x = 3(15) = \textbf{45}$$

ANSWER: m$\angle EBC = 15°$, m$\angle DBE = 30°$, and m$\angle ABD = 45°$

For Example 2, you need to remember that *the sum of the angles in a triangle equals 180°*. Also, an angle in a triangle is often represented by one letter only.

EXAMPLE 2 In $\triangle ABC$, $\angle B = 3\angle A$, and $\angle C = 5\angle A$.

What is the measure of each angle?

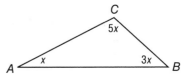

STEP 1 Let m$\angle A = x$, m$\angle B = 3x$, and m$\angle C = 5x$.

STEP 2 Set the sum of the angles equal to 180° and solve the equation.

$$x + 3x + 5x = 180$$
$$9x = 180$$
$$x = \frac{180}{9} = \textbf{20}$$
$$3x = 3(20) = \textbf{60}$$
$$5x = 5(20) = \textbf{100}$$

ANSWER: m$\angle A = 20°$, m$\angle B = 60°$, and m$\angle C = 100°$

For each problem, solve for the angles as indicated.

1. $m\angle ABC = 90°$
   $\angle DBE = 3\angle EBC$
   $\angle ABD = 5\angle EBC$

   What is the measure of each angle?

   $m\angle EBC =$
   $m\angle DBE =$
   $m\angle ABD =$

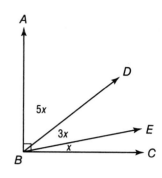

2. $m\angle D + m\angle E + m\angle F = 180°$
   $\angle E = 3\angle D$
   $\angle F = 2\angle D$

   What is the measure of each angle?

   $m\angle D =$
   $m\angle E =$
   $m\angle F =$

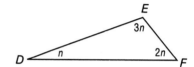

3. $m\angle ABC = 90°$
   $\angle DBE = 5\angle EBC$
   $\angle ABD = 9\angle EBC$

   What is the measure of each angle?

   $m\angle EBC =$
   $m\angle DBE =$
   $m\angle ABD =$

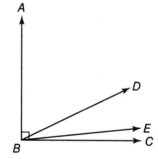

4. $m\angle A + m\angle B + m\angle C = 180°$
   $\angle B = 4\angle A$
   $\angle C = 5\angle A$

   What is the measure of each angle?

   $m\angle A =$
   $m\angle B =$
   $m\angle C =$

Right Triangles and the Pythagorean Theorem

A **right triangle** is a triangle in which two sides meet at a right angle (90°). The side opposite the right angle is called the **hypotenuse.**

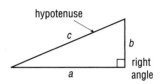

The Greek mathematician Pythagoras discovered that the square of the length of the hypotenuse of a right triangle is equal to the sum of the squares of the lengths of the other two sides. This relation is called the **Pythagorean theorem.**

Using the labels in the triangle above, you can write the Pythagorean theorem.

Pythagorean theorem: $c^2 = a^2 + b^2$

In words: hypotenuse squared = side squared + side squared

EXAMPLE 1 Find the length of the hypotenuse in the triangle at the right.

STEP 1 Substitute 3 for a and 4 for b in the Pythagorean theorem.

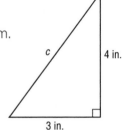

$$c^2 = a^2 + b^2$$
$$c^2 = 3^2 + 4^2$$
$$= 9 + 16$$
$$= 25$$

STEP 2 Solve for c.

$$c^2 = 25$$
$$c = \sqrt{25}$$
$$c = 5$$

ANSWER: **length of hypotenuse = 5 inches**

EXAMPLE 2 Find the length of the unlabeled side in the triangle at the right.

STEP 1 Substitute 10 for c and 6 for a in the Pythagorean theorem.

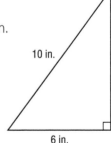

$$c^2 = a^2 + b^2$$
$$10^2 = 6^2 + b^2$$
$$100 = 36 + b^2$$

STEP 2 Subtract 36 from each side and solve for b.

$$100 - 36 = b^2$$
$$64 = b^2$$
$$\sqrt{64} = b$$
$$8 = b$$

ANSWER: **length of unlabeled side = 8 inches**

Solve.

1. The two sides of a right triangle measure 6 feet and 8 feet. What is the length of the hypotenuse?

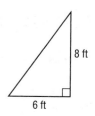

2. The hypotenuse of a right triangle is 15 inches long. If one side measures 12 inches, what is the length of the other side?

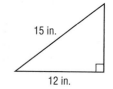

3. A ladder leans against the side of a house. The length of the ladder is 17 feet. The base of the ladder is 8 feet from the house. How far off the ground is the top of the ladder?

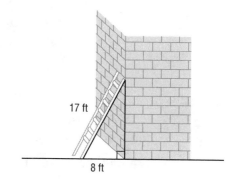

4. Bill has a garden in the shape of a right triangle. One side measures 5 yards and the other side measures 12 yards. How long is the third side—the side opposite the right angle?

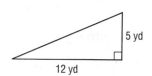

5. In a right triangle, one side measures 5 feet and the second side measures 7 feet. Find the approximate length of the hypotenuse.

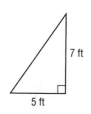

6. A ship sails 10 miles west and 7 miles north of the harbor. Approximately how far is the ship from the harbor?

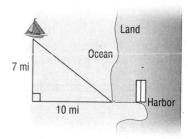

Using Algebra in Trigonometry

Trigonometry is the study of the relations of the sides and angles of right triangles. There are **trigonometric functions** for each angle. Each function is a ratio of two sides of the triangle. The three most important trigonometric functions are shown below.

Trigonometric Functions for $\angle\alpha$

Right Triangle Containing $\angle\alpha$

Sine $\angle\alpha = \frac{a}{c}$

Think of Sine as *opposite side ÷ hypotenuse*

Cosine $\angle\alpha = \frac{b}{c}$

Think of Cosine as *near side ÷ hypotenuse*

Tangent $\angle\alpha = \frac{a}{b}$

Think of Tangent as *opposite side ÷ near side*

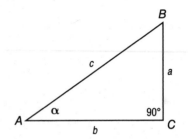

EXAMPLE 1 $\triangle DEF$ is a right triangle with approximate side lengths as indicated. $\angle D = 45°$.

What is the value of sine $\angle D$ and tangent $\angle D$?

To find these values, write the ratio of each pair of indicated sides:

ANSWER: **Sine $\angle D = \frac{15}{21}$, or about 0.7**

Tangent $\angle D = \frac{15}{15} = 1$

Algebra is often used to solve problems in trigonometry.

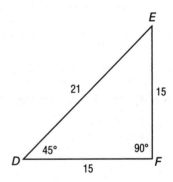

EXAMPLE 2 $\triangle XYZ$ is a right triangle. If sine $30° = \frac{1}{2}$, what is the length h of the hypotenuse of $\triangle XYZ$?

STEP 1 Write sine 30° as a ratio of sides:

Sine $30° = \frac{1}{2} = \frac{10}{h}$

STEP 2 Solve the equation $\frac{1}{2} = \frac{10}{h}$ for h.

Multiply by h: $\frac{h}{2} = 10$

Multiply by 2: $h = 10 \times 2$

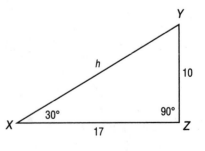

ANSWER: $h = 20$

To solve trigonometry problems you often need to know the value of sine, cosine, and tangent for a variety of angles. Many calculators can find these values for you. On this page you can use the values given in the **Table of Trigonometric Values** below.

Table of Trigonometric Values

Angle	30°	45°	60°
Cosine	0.87	0.71	0.50
Sine	0.50	0.71	0.87
Tangent	0.58	1.00	1.73

1. A forester is 250 feet from a tall fir tree. He finds that the top of the tree is at a 30° angle of elevation from where he stands. The height of the tree is labeled as h.

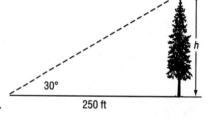

   a. Using the drawing, find the tangent of the angle of elevation.

   Tangent 30° = _____ (as a ratio)

   b. Using the Table of Trigonometric Values, find the tangent of 30°.

   Tangent 30° = _____ (as a number)

   c. Using your answers from parts a and b, determine the height h.

2. Using radar, the Coast Guard determines that a ship is 2.4 miles from the lighthouse. They also find that the ship is 60° west of due north.

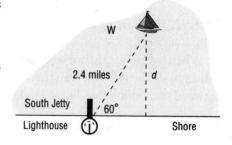

   a. Use the Table of Trigonometric Values to find the sine of 60°.

   b. Use your answer from part a to find d.

3. An airplane flies in a northeast direction at a speed of 200 miles per hour.

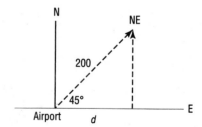

   a. Using the Table of Trigonometric Values, determine cosine 45°.

   b. Using the value found in part a, determine how far directly east of the airport the plane is after 1 hour of flight.

Finding the Distance Between Points on a Grid

You may be asked to find the distance between two points on a grid. If the points lie on a horizontal or vertical line, simply find the number of spaces between the points.

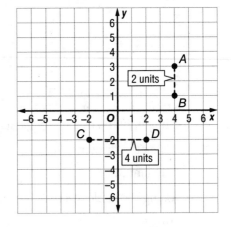

EXAMPLE 1 On the grid at the right, what is the distance between point *A* and point *B?*

     STEP 1 Notice that the points lie on a vertical line parallel to the *y*-axis.

     STEP 2 Count the number of units between the *y*-coordinates. 3 − 1 = **2 units**

ANSWER: 2 units

EXAMPLE 2 Find the distance between point *C* and point *D*.

     STEP 1 Notice that the points lie on a line parallel to the *x*-axis.

     STEP 2 Count the number of units between the *x*-coordinates.
2 − (−2) = 2 + 2 = **4 units**

ANSWER: 4 units

When the points do not lie on a horizontal or vertical line, they lie on a line called a **diagonal.** The Pythagorean theorem (earlier lesson) is used to find the distance between two points that lie on a diagonal.

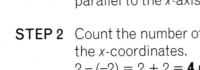

EXAMPLE 3 Find the distance between point *E* and point *G*.

     STEP 1 Draw three lines to connect points *E, F,* and *G*.

     STEP 2 Find the difference between the *x*-coordinates and the difference between the *y*-coordinates.

              (Subtract the lesser from the greater.)

              x-coordinates: 5 − (−1) = 5 + 1 = **6**

              y-coordinates: 5 − (−3) = 5 + 3 = **8**

              6 and 8 are the lengths of the two shorter sides.

STEP 3 Substitute the values 6 and 8 into the Pythagorean theorem and solve for *EG*—the diagonal from point *E* to point *G* and the hypotenuse of right triangle *EFG*.

$$c^2 = a^2 + b^2$$
$$(EG)^2 = 6^2 + 8^2$$
$$= 3^6 + 6^4$$
$$= 100$$
$$EG = \sqrt{100}$$
$$\boldsymbol{EG = 10}$$

ANSWER: 10 units

Find the distance between the points on the grid.

1. points *A* and *B*

2. points *A* and *C*

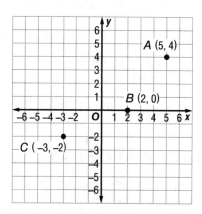

Find the distance between the points on the graphed lines.

3. points *A* and *B*

4. points *B* and *C*

5. points *A* and *C*

6. points *A* and *D*

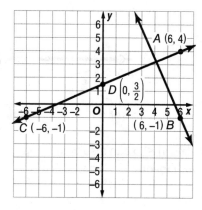

Absolute Value

Absolute value is the distance from zero on a number line. It is the positive value of a number or expression regardless of its sign. The symbol for absolute value is $|\ |$.

Number	Absolute Value	Read as
7	$|7| = 7$	The absolute value of 7 is 7.
−7	$|-7| = 7$	The absolute value of −7 is 7.

EXAMPLE 1 If $|x| = 3$, what are the possible values of x?

Notice that $|3| = 3$ and $|-3| = 3$.

ANSWER: $x = 3$ and $x = -3$

On a number line, the distance between the numbers x and a is equal to $|x - a|$. An unknown distance may be written as an **absolute-value equation.**

EXAMPLE 2 If $|n - 3| = 2$, what are the possible values of n?

The same question can be asked this way. "On a number line, what values of n are each 2 units away from 3?"

1 is 2 units to the left of 3.   5 is 2 units to the right of 3.
$|1 - 3| = |-2| = 2$ $|5 - 3| = |2| = 2$

ANSWER: $n = 1$ and $n = 5$

To tell a range of values, an **absolute-value inequality** such as $|x - 1| < 6$ may be used.

EXAMPLE 3 For what values of x is $|x - 1| < 6$ true?

$|x - 1| < 6$ is true for all values of x that are *less than 6 units away from 1*. The graph shows this range of values.

ANSWER:

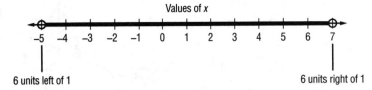

(**Note:** $|x - 1| < 6$ can also be written as an *and* inequality (see page 131). As an *and* inequality, the values of x are written $-5 < x < 7$.)

Write the absolute value of each number.

1. $|-4| =$ $|0| =$ $|8.5| =$ $|-\frac{1}{2}| =$ $|+9| =$

Find the value of each expression. The first one is done as an example.

2. $|7-9| = |-2| = 2$ $|3-12| =$ $|-1-6| =$

Find the solutions to each absolute-value equation.

3. $|x| = 5$ $|n| = 3$ $|r| = 10$ $|w| = 0$ $|z| = 100$

4. $|x-2| = 3$ $|n-5| = 1$ $|r-7| = 15$

Graph each absolute-value inequality.

5. $|x-4| < 5$

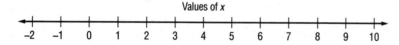

Values of x

6. $|n-1| \leq 3$

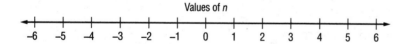

Values of n

7. Write an absolute-value inequality to represent the values of r.

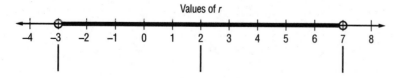

Values of r

Algebra Posttest B

This posttest gives you a chance to check your algebra skills in a multiple-choice format, as used in the GED test and other standardized tests. Take your time and work each problem carefully before you choose your answer. When you finish, check your answers with the answer key at the back of the book.

1. Which addition is shown by the number arrows?

   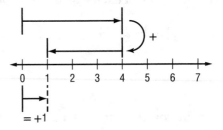

   a. $0 + 4 = 4$

   b. $4 + 1 = 5$

   c. $3 + (-4) = -1$

   d. $4 + (-1) = +3$

   e. $4 + (-3) = +1$

2. At 7:00 P.M., the temperature in Anchorage, Alaska, was –9°F. By 3:00 A.M., the temperature had dropped 15°F. What was the temperature at 3:00 A.M.?

   a. –6°F   **b.** –15°F   **c.** –24°F   **d.** +6°F   **e.** +9°F

3. What is the sum of the product $(-3)(-7)$ and the quotient $\frac{(-10)}{2}$?

   a. 26   **b.** 16   **c.** 0   **d.** –16   **e.** –26

4. What is the value of $(-2)^4$?

   a. –16   **b.** –8   **c.** –4   **d.** 8   **e.** 16

5. Which power equals the quotient $\frac{6^8}{6^3}$?

   a. 6^5 **b.** 6^{11} **c.** 6^{24} **d.** 6^{-5} **e.** 6^{-11}

6. Find the square root. $\sqrt{\frac{36}{25}}$

    a. $\pm\frac{12}{5}$ **b.** $\pm\frac{6}{5}$ **c.** $\pm\frac{6}{25}$ **d.** $\pm\frac{25}{6}$ **e.** $\pm\frac{5}{6}$

7. Which algebraic expression is read "4 times the quantity n minus 7"?

    a. $4n - 7$ **b.** $4 - 7n$ **c.** $4(n - 7)$ **d.** $4(7 - n)$ **e.** $4(-7n)$

8. Evaluate the algebraic expression $(a^2 + 1)(b^2 - 1)$ when $a = 2$ and $b = 3$.

    a. 28   **b.** 32   **c.** 36   **d.** 40   **e.** 44

9. Using the formula $°C = \frac{5}{9}(°F - 32)$, what is the Celsius temperature ($°C$) when the Fahrenheit temperature ($°F$) is $50°F$?

    a. $10°C$ **b.** $18°C$ **c.** $32°C$ **d.** $100°C$ **e.** $212°C$

10. Which equation is *not* represented by the drawing?

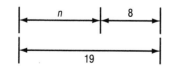

    a. $19 - n = 8$ **b.** $n + 8 = 19$ **c.** $n - 8 = 19$ **d.** $19 - 8 = n$ **e.** $8 + n = 19$

11. What is the first step in solving the equation $y - 8 = 23$?

    a. Subtract $+8$ from each side of the equation.
    b. Multiply each side of the equation by $+8$.
    c. Divide each side of the equation by $+8$.
    d. Add -8 to each side of the equation.
    e. Add $+8$ to each side of the equation.

12. What is the first step in solving the equation $3x + 9 = 15$?

    a. Divide 15 by 3.
    b. Add 9 to each side of the equation.
    c. Multiply each side of the equation by 3.
    d. Subtract 9 from each side of the equation.
    e. Subtract $2x$ from each side of the equation.

13. What is the solution to the equation $3x - 8 = 46$?

    a. $x = 18$ **b.** $x = 16$ **c.** $x = 14$ **d.** $x = 12$ **e.** $x = 10$

14. What is the solution to the equation $2(n + 4) = n + 12$?

    a. $n = 0$ **b.** $n = 2$ **c.** $n = 4$ **d.** $n = 6$ **e.** $n = 8$

15. Kim, Yvette, and Jeff have a house painting business. Kim earns $40 less a month than Yvette. Jeff earns twice as much as Kim. During June, the business earned $2,620. Which equation can be used to find the dollar amount x that Kim earned during June?

    a. $(x + \$40) + 2x = \$2,620$
    b. $(x - \$40) + 2x = \$2,620$
    c. $x + (x - \$40) + 2x = \$2,620$
    d. $x + (x + \$40) + 2x = \$2,620$
    e. $2x + (x + \$40) - x = \$2,620$

16. The ratio of two numbers is 5 to 4. The sum of the numbers is 99. What number is the greater of the two numbers?

    a. 33   **b.** 44   **c.** 55   **d.** 66   **e.** 77

17. For how many different values of x is the equation $3x^2 = y$ true when $y = 75$?

    a. none   **b.** one   **c.** two   **d.** three   **e.** four

18. Jani makes punch by mixing 3 cups of pineapple juice with 5 cups of orange juice. Which proportion tells the number n of cups of pineapple juice Jani should mix with 18 cups of orange juice?

    a. $\dfrac{n}{3} = \dfrac{5}{18}$ **b.** $\dfrac{n}{5} = \dfrac{3}{18}$ **c.** $\dfrac{n}{18} = \dfrac{5}{8}$ **d.** $\dfrac{n}{18} = \dfrac{3}{5}$ **e.** $\dfrac{n}{18} = \dfrac{3}{8}$

For problems 19 and 20, refer to the grid at the right.

19. Which point is the x-intercept of line L?

    a. $(0, 4)$
    b. $(2, 2)$
    c. $(0, 0)$
    d. $(6, -2)$
    e. $(4, 0)$

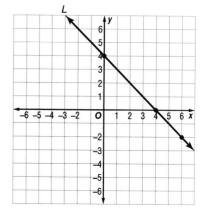

20. What is the slope of line L?

$$\text{slope} = \frac{\text{change in } y}{\text{change in } x}$$

    a. -2
    b. -1
    c. 0
    d. 1
    e. 2

21. Which point does *not* lie on the line of solutions of the graph of the linear equation $y = 2x - 3$?

   a. $(-1, -5)$ **b.** $(0, -3)$ **c.** $(1, 2)$ **d.** $(2, 1)$ **e.** $(3, 3)$

22. Which number is *not* an allowed value for x in the inequality $2x - 6 \leq 5$?

   a. -6 **b.** -4 **c.** 0 **d.** 4 **e.** 6

23. Solve the inequality $3 - 4x > -1$.

   a. $x < \dfrac{1}{2}$ **b.** $x > \dfrac{1}{2}$ **c.** $x > 1$ **d.** $x < 1$ **e.** $x > 4$

24. Which inequality is graphed on the number line?

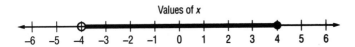

Values of x

   a. $-4 \leq x + 4$
   b. $-4 < x \leq 4$
   c. $-4 < x < 4$
   d. $-4 \leq x \leq 4$
   e. $-4 \leq x < 4$

25. Subtract: $(6n^2 - n) - (3n^2 + 2n)$

   a. $3n^2 - 3n$
   b. $3n^2 + 3n$
   c. $3n^2 + n$
   d. $9n^2 + n$
   e. $9n^2 - 3n$

26. Divide: $\dfrac{(26x^5 + 16x^3)}{2x^3}$

   a. $13x^2 + 8x^3$
   b. $13x^2 + 16x^3$
   c. $13x^2 + 8x$
   d. $13x^2 + 16x$
   e. $13x^2 + 8$

27. What is the volume of the rectangular solid at the right?

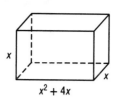

$x^2 + 4x$

   a. $x^2 - 2x$
   b. $x^2 + 2x$
   c. $x^4 - 4x^3$
   d. $x^4 + 4x^3$
   e. $x^4 - 4x$

28. Simplify: $\sqrt{49x^6 y}$

   a. $7x^3\sqrt{y}$ **b.** $7x^3\sqrt{3y}$ **c.** $7x^2\sqrt{y}$ **d.** $7x\sqrt{y}$ **e.** $7x\sqrt{3y}$

29. Factor the expression $16n^4 - 24n^2$ as completely as possible.

   a. $n^2(16n^2 - 24)$
   b. $n^2(16n^2 - 48)$
   c. $4n^2(4n^2 - 6)$
   d. $8n^2(2n^2 - 3)$
   e. $8n^2(2n^2 - 3n)$

30. Simplify the quotient $\frac{r^2 - 9}{r + 3}$.

   a. $r^2 + 3$
   b. $r^2 - 3$
   c. $3r^2$
   d. $r + 3$
   e. $r - 3$

ALGEBRA POSTTEST B CHART

Problem Numbers	Skill Area	Practice Pages
1, 2, 3	signed numbers	10–23
4, 5, 6	powers and roots	26–39
7, 8, 9	algebraic expressions	42–53
10, 11	one-step equations	56–67
12, 13, 14, 15	multistep equations	72–87
16, 17, 18	ratios, proportions, and quadratics	92–105
19, 20, 21	graphing equations	108–121
22, 23, 24	inequalities	124–133
25, 26, 27	polynomials	136–149
28, 29, 30	factoring	152–161

ANSWER KEY

Pages 1–6

1. $-1 + -6 = -7$
2. $5 + -3 = 2$
3. 11
4. -6
5. -30
6. 2
7. 16
8. -27
9. 6^6
10. 7^3
11. 8
12. $\frac{5}{7}$
13. $3(n + 6)$
14. 20

15. 16
16. $100°C$
17. **c.** $x - 8 = 21$
18. **a.** $x = 17$
    b. $y = 22$
19. **a.** $n = 6$
    b. $n = 36$
20. **a.** $x = 9$
    b. $s = 48$
21. **a.** $y = 8$
    b. $n = 8$
22. **a.** $x = 13$
    b. $b = 5$
23. 28 and 21
24. $5\frac{1}{3}$ cups
25. $x = 5$ or $x = -5$

26.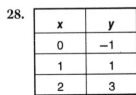

27. slope $= \frac{4}{3}$

28.
x	y
0	−1
1	1
2	3

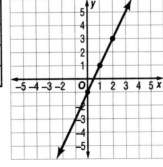

29. $n > -3$
30. $n \geq 2$

31. $x < 7$

32.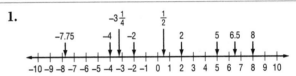

33. $-3x^2$ and $5x^2$
34. $3x^2 - x$
35. $a^2 - 5b$
36. $6n^3 + 8n^2 - 4n$
37. $7x^2 + 3x$
38. $5\sqrt{3}$
39. $6x^2\sqrt{y}$
40. $2(2x^2 + 5)$
41. $4z^2(3z + 4)$
42. $3r^2 + 4$
43. 7

Page 10

1. $+8, +3, +12, +7$
2. $-5, -4, -19, -8$
3. $+2\frac{1}{2}, +4.75, -5\frac{2}{3}, -\frac{1}{4}$
4. positive, negative, positive, negative
5. positive, positive, negative, negative

Page 11

1.

2.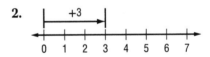

3.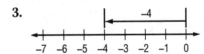

Pages 13 and 15

1. $+4 + (+3) = +7$
2. $-1 + (-6) = -7$
3. $+2 + (+4) = +6$
4. $-2 + (-3) = -5$
5.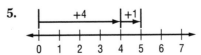

6.

7.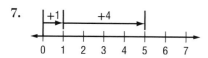

8.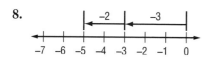

For problems 9–11, a + sign does not need to be placed before answers that are positive.

9. 11, 11, 19, 16

10. −7, −7, −7, −12

11. −21, −21, −25, −41

12. 5 + (−3) = +2

13. 5 + (−5) = 0

14. 1 + (−3) = −2

15. −2 + 4 = +2

16.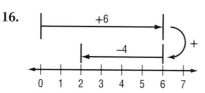

17.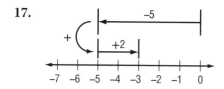

18. 3, −3, 0, −1

19. 1, −3, 4, −7

20. 0, −5, 5, 9

Page 17

1. b. 5 − (−4)

2. a. 2 − (−3)

For problems 3–6, a + sign does not need to be placed before answers that are positive.

3. 6, 5, 4, 8

4. −4, 5, −4, 6

5. 5, 10, 18, 23

6. −6, −12, −1, −6

7. 13°

8. 111° below freezing

9. 20,600 feet

10. 14,775 feet

Page 19

A + sign does not need to be placed before answers that are positive.

1. 18, −35, 32, 54

2. 56, 20, −48, −12

3. 36, 28, 42, 18

4. −40, −9, −36, −30

5. 45, −54, 21, −24

6. 30, 16, 48

7. −24, −18, −36

8. 12, −40, 48

9. −16, 30, 8

Page 21

A + sign does not need to be placed before answers that are positive.

1. −6, −5, 6, 7

2. 13, −34, 18, −14

3. −3, −9, 9, 8

4. −25, −14, 30, −24

5. −2, −2, 6, −4

6. −14, −15, 12, −25

7. $-\frac{2}{3}, -\frac{3}{4}, -\frac{1}{2}, -\frac{2}{5}$

8. $\frac{1}{2}, \frac{2}{5}, \frac{3}{4}, \frac{2}{3}$

9. $-\frac{1}{5}, -\frac{2}{5}, -\frac{7}{20}, -\frac{2}{3}$

Pages 22–23

1. −3.0%

2. 480 feet

3. a. 2,400 feet below sea level
   b. −2,400 feet (elevation)

4. 46.73

5. $818.50 owed

6. −13° change

7. **a.** depth is 210 feet below the surface
   b. −210 feet

8. −1 foot

9. −24°

10. 300 more truckloads

11. **a.** 120 feet
    b. 3 to 1, or $\frac{3}{1}$

12. **a.** 18 pounds lost
    b. −3 lb, or 3 lb lost per month

Pages 24–25

1. $+13, +2\frac{1}{2}, -26, -\frac{7}{8}$

2.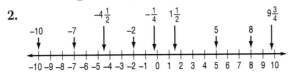

3. $+5 + (-3) = +2$

4. $-2 + (+5) = +3$

For problems 5–8 and 10–16, a + sign does not need to be placed before answers that are positive.

5. 5, 3, 0, 9

6. −12, −12, 3, −4

7. 5, 0, −3, 19

8. 2, −13, −8, −6

9. 23°

10. −27, −28, −18, −16

11. 30, 15, 32, 54

12. −24, 36, 18

13. −5, −7, 8, 8

14. −4, 8, −7, −6

15. −6, −2, 3, −9

16. $\frac{7}{8}, -\frac{4}{5}, -\frac{3}{4}, \frac{2}{3}$

Pages 27–28

1. $(-1)^2, -1$ to the second power or −1 squared

2. $5^2, 5$ squared or 5 to the second power

3. $\left(\frac{1}{2}\right)^2, \frac{1}{2}$ squared or $\frac{1}{2}$ to the second power

4. $6^2, 6$ squared or 6 to the second power

5. $7^3, 7$ cubed or 7 to the third power

6. $(-3)^3, -3$ cubed or −3 to the third power

7. $6^3, 6$ cubed or 6 to the third power

8. $\left(-\frac{3}{4}\right)^3, -\frac{3}{4}$ cubed or $-\frac{3}{4}$ to the third power

9. $2^4, 2$ to the fourth power

10. $5^4, 5$ to the fourth power

11. $(-4)^4, -4$ to the fourth power

12. $\left(\frac{1}{4}\right)^4, \frac{1}{4}$ to the fourth power

13. 25, 4, 81, 36

14. 125, 81, −1000, 81

15. $\frac{4}{9}, -\frac{64}{125}, \frac{1}{16}, \frac{27}{64}$

16. 61, 80, 5

17. 14, 56, 21

18. $\frac{3}{4}, \frac{1}{16}, 1\frac{1}{16}$

Page 29

1. 128, 432, −200

2. $3\frac{1}{8}, 5, \frac{4}{27}$

3. $2\frac{27}{49}, 1\frac{32}{49}, -4$

4. $3\frac{5}{9}, 2\frac{13}{18}, \frac{5}{18}$

Pages 30–31

1. $2^7, 3^4, 4^7$

2. $5^3, 7^4, 4^4$

3. $6^3, 2^5, 8^7$

4. $4^5 \cdot 3^4, 5^{12} \cdot 2^3, 7^6 \cdot 5^3$

5. $5^2, 2^4, 8^2$

6. $7^1 = 7, 8^1 = 8, 5^1 = 5$

7. $4^2, 8^3, 7^1 = 7$

8. $6^3 \cdot 4^2, 10^2 \cdot 5^4, 9^2 \cdot 7^3$

Page 32

1. $\frac{1}{7 \times 7}$, 7 to the negative 2nd power, $\frac{1}{49}$
2. $\frac{1}{10 \times 10 \times 10}$, 10 to the negative 3rd power, $\frac{1}{1,000}$
3. $\frac{1}{3 \times 3 \times 3 \times 3}$, 3 to the negative 4th power, $\frac{1}{81}$
4. 3^2, 4^0 or 1, 10^{-1}
5. 6^3, 2^6, 5^4

Page 33

1. 630; 8,750,000; 90,000,000
2. 0.051; 0.000085; 0.000007
3. 5×10^3, 3×10^4, 7.5×10^7
4. 3×10^{-2}, 7.5×10^{-3}, 1.25×10^{-5}
5. 240,000 miles
6. 1.2×10^{-3} inches

Page 35

1. ± 13, ± 5, ± 9
2. ± 2, ± 11, ± 6
3. ± 7, ± 1, ± 15
4. ± 12, ± 8, ± 3
5. ± 10, ± 4, ± 14
6. $\pm \frac{2}{3}$, $\pm \frac{3}{8}$, $\pm \frac{5}{6}$
7. $\pm \frac{1}{2}$, $\pm \frac{7}{9}$, $\pm \frac{11}{12}$
8. $\pm \frac{9}{4}$, $\pm \frac{1}{14}$, $\pm \frac{10}{11}$

Page 37

1. 6.5
2. 4.4
3. 8.7
4. 9.5
5. 5.3
6. 7.1
7. 11.7
8. 14.4

Pages 38–39

1. 64
2. a. 16 feet
   b. 64 feet
3. a. 0.1 cubic inches
   b. about 10^4 cubic inches
4. $3^2 - 2^3 = 9 - 8 = 1$
5. 4.2×10^4 grams = 42,000 grams
6. 10^7 atoms

7. a. 0.006 inch
   b. 6×10^{-3} inch
   c. 1.2×10^{-1} inch
8. 4^5 trees
9. 3^3, or 27
10. a. about 4.9×10^2 seconds
    b. about 8 minutes
11. 10 bananas
12. about 12.2 feet

Pages 40–41

1. 3 is the base, 4 is the exponent
2. $(-4)^4$
3. 5^3
4. $\left(-\frac{3}{5}\right)^2$
5. 8^2
6. $\left(\frac{2}{3}\right)^3$
7. 9^4
8. 81
9. -125
10. 81
11. 1
12. 7
13. $\frac{4}{9}$
14. 17
15. $1\frac{1}{2}$
16. $7\frac{1}{9}$
17. $\frac{9}{10}$
18. 5^9
19. 8^3
20. $\frac{1}{64}$
21. 2
22. $\frac{1}{1,000}$
23. 45,000
24. 0.0023
25. 7.2×10^5
26. 6×10^{-3}
27. ± 6
28. $\pm \frac{7}{8}$
29. $\sqrt{56} \approx 7.5$
30. $\sqrt{83} \approx 9.1$

Pages 42–43

For problems 1–12, answers may vary.

1. $n - 8$
2. $\frac{s}{5}$
3. $y + 12$
4. $w - 4$
5. $7x$
6. $19 + r$
7. 5 times x
8. 4 plus n
9. 7 minus n
10. m divided by 2
11. a divided by 5
12. s subtract 3
13. e
14. h

15. a

16. f

17. g

18. b

19. c

20. d

Pages 44–45

1. c. $h + 144$

2. a. $\frac{t}{3}$

3. d. $x - \$2.95$

4. b. $\frac{(n-5)}{3}$

5. d. $n - 0.05n$

6. c. $g^2 - 6$

7. a. $6n^2 + 3$

8. c. $(n-3)^2$

For problems 9–14, answers may vary.

9. $h + 154$

10. $\frac{n}{5}$

11. $0.3p$

12. $m - 0.25m$

13. $3n^2 - 4$

14. $2b^2 - 12b$

Pages 47–48

1. 5 -9 4

2. 20 -14 16

3. 7 -6 5

4. -8 6 -8

5. 57   0   13

6. 9 -39 18

7. 27   8 -7

8. 27   100   21

9. 10   12

10. -20 $\frac{1}{3}$

Pages 49–50

1. 22 feet

2. 28 feet

3. 44 inches

4. 84 square feet

5. 42 square inches

6. 616 square inches

7. 720 cubic inches

8. 88 cubic feet

Page 51

1. 18 feet

2. 466 yards

3. about 176 feet

4. 180 square feet

5. 18 square yards

6. about 1,385 square meters

7. 756 cubic feet

Page 53

1. -4 and x; 3 and a; -5 and z

2. 1 and y^2; 1 and z^2; -1 and a^2

3. -9 and x^2y; -4 and c^2d; 7 and uv^2

4. 3 and a^2b^2; 8 and x^2y^2; -7 and ab^3

5. x and $-5x$; $-y^3$ and $-2y^3$; a^2b and $2a^2b$

6. a^2 and $-4a^2$; no like terms; x^3 and $4x^3$

7. c^3 and $-5c^3$, $-rs^2$ and $3rs^2$; $-4x^3y$ and $2x^3y$

8. $2x + 7$

9. $13y + 12$

10. $3s^2 + s + 9$

11. $10x + 8$

12. $4y^2 + 6$

Pages 54–55

For problems 1–16, answers may vary.

1. $x - 7$

2. $\frac{z}{9}$

3. $y + 19$

4. $-4n$

5. $6(x + 4)$

6. $2(y - 5)^2$

7. 12 plus y

8. x minus 9

9. 15 times a

10. y divided by 12

11. 4 times s plus 8

12. -9 times the quantity 2 times x plus 4

13. $6s$

14. $m - \$12.49$

15. $\frac{(n-2)}{6}$

16. $n - 0.4n$

17. -2

18. 34

19. 7

20. 60

21. -9

22. -16

23. 12

24. 4

25. about 283 cubic feet

26. $74.25

27. 180 square feet

28. 40°C

Pages 56–57

1. b. $9n = 14$

2. c. $\frac{n}{6} = 50$

For problems 3–10, answers may vary.

3. $y + 9 = 26$

4. $x + 14 = 21$

5. $t - 13 = 15$

6. $x - 21 = 43$

7. $5x = 60$

8. $4y = 24$

9. $\frac{n}{3} = 7$

10. $\frac{s}{6} = 2$

Page 58

1. No.

2. Yes.

3. Yes.

4. No.

5. Yes.

6. No.

7. Yes.

8. Yes.

Page 60

1. $x = 3, x = 5, y = 6, n = 14$

2. $y = 13, s = 5, x = 23, z = 18$

3. $x = 3\frac{1}{2}, y = 2\frac{1}{3}, c = \frac{1}{2}, r = 4\frac{1}{4}$

4. $x = 3.6, n = 3.3, z = 8.5, y = 4.75$

Page 61

1. $x = 20, x = 10, y = 16, z = 47$

2. $y = 17, x = 26, y = 37, n = 36$

3. $x = 6\frac{2}{3}, r = 3\frac{5}{8}, y = 2, v = 8$

4. $m = 7.7, x = 8.1, n = 17.25, y = 11.75$

Page 62

1. c. $n - 18 = 14, n = 32$

2. c. $t + \$125.00 = \$435.75, t = \$310.75$

3. d. $p + 0.25(\$40) = \$40, p = \$30$

4. c. $t + 15°F = 87°F, t = 72°F$

For problems 5 and 6, equations may vary.

5. $p - \$14.95 = \$34.00, p = \$48.95$

6. $n - 14 = 19, n = 33$

Page 63

1. $x = 5, n = 6, y = 9, m = 7$

2. $y = -6, x = -9, b = -13, y = -11$

3. $x = 3, n = 7, z = 21, y = 11$

4. $y = 2.4, y = 6.5, x = 5.4, n = 7.5$

Page 64

1. $x = 20, y = 27, c = 88, n = 48$

2. $y = 18, x = 68, n = 36, t = 180$

3. $x = 7, y = 7, n = 15, r = 13.5$

Page 65

1. c. $14n = 168, n = 12$

2. b. $0.05w = 9, w = 180$

3. a. $\frac{n}{4} = 7, n = 28$

4. b. $p = \frac{\$16.50}{3}, p = \$5.50$

For problems 5 and 6, equations may vary.

5. $1.5h = 42, h = 28$

6. $\frac{s}{8} = 4, s = 32$

Page 66

1. $x = 8, z = 12, a = 28, n = 12$

2. $n = -\frac{7}{4}, s = -1, x = -\frac{8}{9}, y = -\frac{35}{16}$

3. $z = 5.25, x = 1.60, c = 17.60, r = 2.10$

Page 67

1. b. $\frac{2}{3}x = 36, x = 54$

2. d. $\frac{1}{3}h = 4, h = 12$

3. c. $\frac{3}{4}n = 18, n = 24$

4. b. $\frac{4}{5}r = \$7.75, r \approx \$9.69$

For problems 5 and 6, equations may vary.

5. $\frac{2}{5}h = 14, h = 35$

6. $\frac{5}{8}c = 50, c = 80$

Pages 68–71

1. $x = 2$

2. $z = 13$

3. $a = -3$

4. $y = 19$

5. $x = 25$

6. $b = -4$

7. $z = 8$

8. $x = -8$

9. $y = -11$

10. $x = 27$

11. $y = 56$

12. $a = -24$

13. $x = -15$

14. $y = 36$

15. $a = -8$

16. $y = 17$

17. $n = 35$

18. $q = 24$

19. $y = 35$

20. $n = -7$

21. $x = -3$

22. $p = 7$

23. $x = 25$

24. $s = 0$

25. $x = \frac{2}{3}$

26. $r = 6.75$

27. $x = \frac{1}{15}$

28. $k = 3.12$

29. $x = 1\frac{1}{5}$

30. $n = 2$

31. $z = 42$

32. $y = 11.25$

33. $x = 4\frac{1}{2}$

For problems 34–39, answers to parts a and b may vary.

34. **a.** $m =$ money Zeta had
   b. $m - \$4.75 = \$2.40$
   c. $m = \$7.15$

35. **a.** $c =$ cookies in each box
   b. $\frac{c}{5} = 6$
   c. $c = 30$

36. **a.** $m =$ money you gave clerk
   b. $m - \$124.77 = \$5.23$
   c. $m = \$130.00$

37. **a.** $n =$ number of students in class
   b. $\frac{3}{4}n = 18$
   c. $n = 24$

38. **a.** $n =$ number of teachers at Grant School
   b. $\frac{1}{6}n = 4$
   c. $n = 24$

39. **a.** $m =$ amount Jocelyn earned during May
   b. $\frac{2}{3}m = \$90$
   c. $m = \$135$

For problems 40–47, equations may vary.

40. $x + 7 = 17, x = 10$

41. $n + 13 = 21, n = 8$

42. $5y = 45, y = 9$

43. $\frac{r}{3} = 4, r = 12$

44. $x + 19 = 28, x = 9$

45. $y + 32 = 62, y = 30$

46. $16w = 64, w = 4$

47. $5l = 80, l = 16$

Page 73

1. $y = 3, x = 3, z = -4$

2. $x = 4, z = 8, a = -7$

3. $y = -8, z = 2, x = -30$

Page 75

1. $x = 6, y = 7, z = 5$

2. $y = 13, a = 5, x = 6$

3. $y = 15, z = -4, n = -24$

4. $x = 1, a = 3, z = 5$

5. $x = 9, z = 21, n = 8$

Pages 76–77

1. $x = 5, y = 4, z = 4$

2. $a = 12, y = 8, x = 5$

3. $x = 4, z = 2, y = 26$

4. $y = 48, a = 50, x = 18$

5. $z = 4, y = 7, a = 2$

6. $a = 10, x = 12, y = 5$

7. $b = 3, z = 2, y = 6$

Pages 78–79

1. $x = 6, y = 3, z = 9$

2. $z = 3, a = 3, x = -3$

3. $x = 8, b = 8, y = 2$

4. $m = 8, x = 2\frac{1}{2}, y = -4$

5. $x = 4, y = 3, z = -9$

6. $a = -10, b = 2, z = 6$

7. $y = 3, x = -5, n = 3$

Pages 80–81

1. **b.** $8n + 9 = 73, n = 8$

2. **c.** $5w - 7 = 3w + 19, w = 13$

3. **d.** $t + 2t = \$2700, t = \$900$

4. **c.** $25a + \$1,850 = \$2,375, a = 21$

For problems 5–12, equations may vary.

5. $3n = n + 12, n = 6$

6. $\frac{2}{3}m + \frac{1}{6}m = 25, m = 30$

7. $p + 0.06p = \$630.70, p = \$595.00$

8. $m + \$50(6) = 3m, m = \$150$

9. $4x + 15 = 27, x = 3$

10. $3n + 16 = 5n + 10, n = 3$

11. $3r = r + 4 + 4 + 2, r = 5$

12. $4y = 2y + 24, y = 12$

13. $x = 8$

14. $x = 16$

Page 83

1. $x = 6, n = 7, b = 5$

2. $y = 11, a = 11, b = 4$

3. $x = 10, m = 7, y = -2$

4. $y = 30, z = 21, a = 13$

Page 85

1. **b.** $3(n + 1) = 2(n + 4), n = 5$

2. **c.** $j + (j - \$60) + 2(j - \$60) = \$4,400, j = \$1,145$

3. $4(n + 2) = 3(n + 5), n = 7$

4. $\frac{2}{3}(y - 3) = \frac{1}{3}y, y = 6$

5. $c + (c + \$4.75) + (\$8.50 + c + \$4.75) = \$72.00$
   Chalk: $18.00
   Erasers: $22.75
   Rulers: $31.25

6. $x + (x + 1) + (x + 2) = 54$
   1st integer: 17
   2nd integer: 18
   3rd integer: 19

Page 87

1. **a.** $r = \frac{C}{2\pi}$
   b. $r \approx 15.6$ feet

2. **a.** $c = P - a - b$
   b. $c = 70$ yards

3. **a.** $h = \frac{V}{lw}$
   b. $h = 8$ feet

4. **a.** $h = \frac{2A}{B}$
   b. $h = 4$ feet

5. **a.** $h = \frac{V}{\pi r^2}$
   b. $h \approx 25.97$ feet

6. $r = \frac{i}{pt}$

Pages 88–91

1. $x = 16$	11. $n = 23$	21. $r = 3$
2. $n = 4$	12. $y = 17$	22. $x = -6$
3. $r = 72$	13. $b = 4$	23. $y = 10$
4. $y = 11$	14. $z = 16$	24. $n = -2$
5. $x = 24$	15. $y = 4$	25. $y = -6$
6. $n = 24$	16. $w = 13$	26. $z = 4$
7. $n = 5$	17. $x = 5$	27. $x = 44$
8. $x = 9$	18. $n = 7$	28. $p = 4$
9. $y = 30$	19. $x = 5$	29. $n = 6$
10. $x = 5$	20. $s = 5$	30. $x = 14$

31. **d.** $6m - 4 = 26, m = 5$

32. **c.** $n + 8(5) = 2n, n = 40$

33. $5(x - 1) = 3(x + 9), x = 16$

34. $4n - 3(n + 2) = 5, n = 11$

35. $c + 24(5) = 3c, c = 60$

36. $h + (h - 8) + 2(h - 8) = 72$
    Alyce: 24 hours
    Jon: 16 hours
    Kim: 32 hours

37. $x + (x + 2) + (x + 4) = 66$
    1st integer: 20
    2nd integer: 22
    3rd integer: 24

38. $(3x - 2) + (2x + 4) + (3x - 3) = 89$, or $8x = 90$,
    $x = 11.25$

For problems 39–42, equations may vary.

39. $3x + 25 = 33, x = 2\frac{2}{3}$

40. $5y + 9 = 4y + 12, y = 3$

41. $3d = d + 9 + 6, d = 7\frac{1}{2}$

42. $3n + 6 = 7n - 26, n = 8$

Page 92

1. $\frac{4}{3}$

2. $\frac{5}{16}$

3. **a.** $\frac{3}{2}$

    b. $\frac{3}{5}$

4. $\frac{2}{7}$

5. $\frac{5}{17}$

Page 93

For problems 1–4, equations may vary.

1. $3x + 2x = 75$, $x = 15$
   1st number $= 45$
   2nd number $= 30$

2. $5x - 2x = 18$, $x = 6$
   1st number $= 30$
   2nd number $= 12$

3. $\$10x + \$12x = \$2{,}420$, $x = \$110$
   You earn $= \$1{,}100$
   Tomas earns $= \$1{,}320$

4. $7x + 4x + 7x + 4x = 176$, $x = 8$
   length $= 56$ feet
   width $= 32$ feet

Page 95

1. No, Yes, No

2. Yes, No, Yes

3. $16n = 16$, $4n = 36$, $20x = 40$, $12x = 30$

4. $5x = 90$, $32n = 96$, $3x = 48$, $12n = 144$

5. $n = 1$, $x = 2$, $x = 3$, $n = 2$

6. $n = 9$, $x = 12$, $n = 8$, $x = 3$

Page 96–97

1. **b.** $\frac{75}{2} = \frac{280}{n}$

2. **c.** $\frac{2}{3} = \frac{g}{19}$

3. **a.** $\frac{450}{8} = \frac{225}{t}$

4. **c.** $\frac{2}{5} = \frac{n}{280}$

For problems 5–10, variables and proportions may vary. However, there is only one correct answer.

5. $c = $ cost of 12 ounces
   $$\frac{2.24}{8} = \frac{c}{12}$$
   $c = \$3.36$

6. $d = $ distance
   $$\frac{d}{16} = \frac{102}{3}$$
   $d = 544$ miles

7. $n = $ parts of brown
   $$\frac{n}{15} = \frac{2}{5}$$
   $n = 6$ parts

8. $n = $ number of centimeters
   $$\frac{n}{2} = \frac{12.7}{5}$$
   $n = 5.08$ centimeters

9. $m = $ additional earnings
   $$\frac{m}{28} = \frac{76.80}{8}$$
   $m = \$268.80$

10. $d = $ drops of hardener
    $$\frac{d}{28} = \frac{3}{7}$$
    $d = 12$ drops

Page 99

1.

x	y
0	4
1	5
2	6
3	7

2.

x	y
1	1
2	4
3	7
4	10

3.

b	a
−1	5
0	3
1	1
2	−1

4.

s	r
−2	2
0	3

5.

w	p
6	18
10	26
15	36
20	46

6.

m∠N	m∠M
25°	65°
45°	45°
62°	28°
78°	12°

Page 101

1. $x = -2, y = -1$
2. $x = -5, y = -9$
3. $x = 1, y = 2$
4. $x = 2, y = 7$
5. $x = -3, y = 3$
6. $x = 4, y = 1$

Page 103

1. $y = 9, y = 8, y = 12$
2. $y = 7, y = 2, y = 37$
3. $x = \pm 5, x = \pm 4, x = \pm 6$
4. $x = \pm 7, x = \pm 8, x = \pm 6$
5. $x = \pm 3, x = \pm 5, x = \pm 13$

Pages 104–105

1. $P = l + 4 + l + 4 = 2l + 8$
2. **c.** $p = (\$9.95 - \$1.25)n - \$120$
3. **a.** \$15,000
   b. \$600
   c. \$500
4. $s = 45, m = 105$
5. $t = \$30, b = \$15$
6. $c = \$8.50, d = \$17.00$
7. about 201 square inches
8. **a.** $h = 32$ feet
   b. a little less than 2.5 seconds
9. \$7,000

Pages 106–107

1. **a.** $\frac{5}{2}$
   b. $\frac{5}{7}$
2. 1st number $= 27$
   2nd number $= 36$
3. length $= 66$ yards
   width $= 22$ yards
4. $n = 6$
5. $y = 2$
6. $x = 32$

For problems 7–8, proportions may vary.

7. $m = \$79.20$
8. $n = 24$
9.

x	y
0	-2
2	0
4	2
6	4

10.

x	y
-1	1
0	3
1	5
2	7

11. $x = -2, y = 1$
12. $x = 0, y = 1$
13. $y = 16$
14. $y = 18$
15. $x = \pm 6$
16. $x = \pm 4$

Pages 108–109

1. **a.** $(4, -2)$
   b. $(-3, 5)$
   c. $(2, 6)$
   d. $(-4, -7)$
2. **a.** $x = 2, y = -8$
   b. $x = -1, y = 6$
   c. $x = 0, y = 5$
   d. $x = -4, y = 0$
3. $A = (-6, 5); B = (-3, 3); C = (-2, 0); D = (2, 4)$
   $E = (0, 2); F = (6, 2); G = (3, 0); H = (5, -2)$
   $I = (1, -3); J = (0, -5); K = (-7, -5); L = (-3, -3)$

Page 110

1.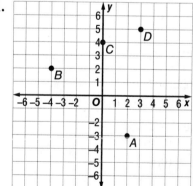

2.

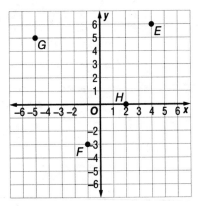

Page 111

1.

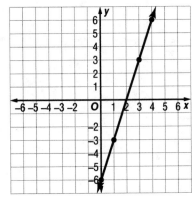

2.

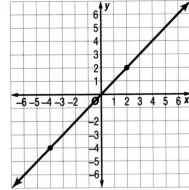

Page 113

1. Line *A*: zero; Line *B*: negative;
   Line *C*: undefined; Line *D*: positive

2. Slope of line $E = 2$
   x-intercept $= (2, 0)$
   y-intercept $= (0, -4)$

3. Slope of line $F = -\frac{4}{3}$
   x-intercept $= (1\frac{1}{2}, 0)$
   y-intercept $= (0, 2)$

4. **a.** 5 **b.** 2 **c.** −1 **d.** $\frac{2}{3}$

Page 114–115

1.

x	y
0	−2
2	0
5	3

x-intercept $= (2, 0)$
y-intercept $= (0, -2)$
slope $= 1$

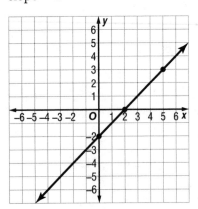

For problems 2 and 3, the Tables of Values will vary from the sample answers given. The graphed lines, intercept values, and slope should be the same as those given.

2.

x	y
0	2
2	4
4	6

x-intercept $= (-2, 0)$
y-intercept $= (0, 2)$
slope $= 1$

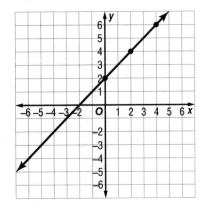

3.

x	y
0	−4
−1	−2
1	−6

x-intercept $= (−2, 0)$
y-intercept $= (0, −4)$
slope $= −2$

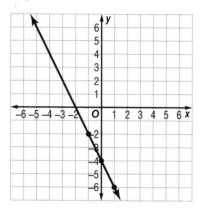

4.

x	y
0	−1
2	$\frac{1}{2}$
4	2

x-intercept $= (\frac{4}{3}, 0)$
y-intercept $= (0, −1)$
slope $= \frac{3}{4}$

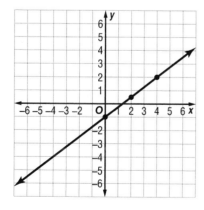

Page 116

For problem 1, Table of Values will vary.

1.

r	C
0	0
4	25.1
10	62.8
15	94.2

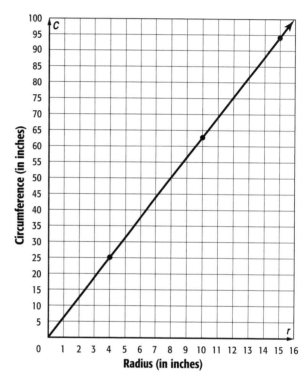

For problems 2–5, estimates will vary.

2. approximately 41 inches

3. approximately 7.5 inches

4. approximately 6.25

Page 117

For problem 1, Table of Values will vary.

1.

°F	°C
10°	−12.2°
32°	0°
50°	10°
100°	≈ 37.8°

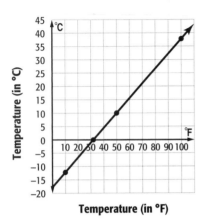

Temperature (in °F)

For problems 2–5, estimates will vary.

2. approximately 18°C

3. 32°F

4. approximately $\frac{1}{2}$ (exact slope is $\frac{5}{9}$)

Pages 118–119

1. $y = 2x + 1$ $y = -x + 4$

x	y
−2	−3
0	1
2	5

x	y
0	4
2	2
4	0

Common solution: $(1, 3)$

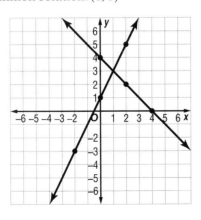

2. $y = 2x + 3$ $y = x + 1$

x	y
−1	1
0	3
1	5

x	y
0	1
2	3
4	5

Common solution: $(-2, -1)$

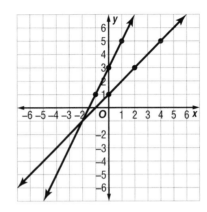

3. $y = 3x + 1$ $y = -x + 5$

x	y
−1	−2
0	1
1	4

x	y
−1	6
0	5
6	−1

Common solution: $(1, 4)$

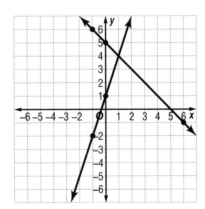

Page 121

1. a.

x	y
−4	11
−3	4
−2	−1
0	−5
2	−1
3	4
4	11

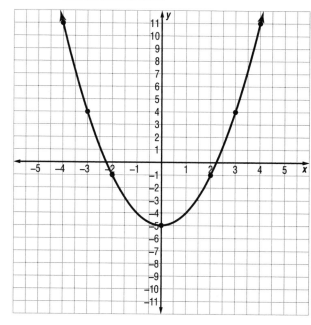

b. approximately $-3\frac{1}{3}$ and $3\frac{1}{3}$

2. a.

x	y
−3	12
−2	2
−1	−4
0	−6
1	−4
2	2
3	12

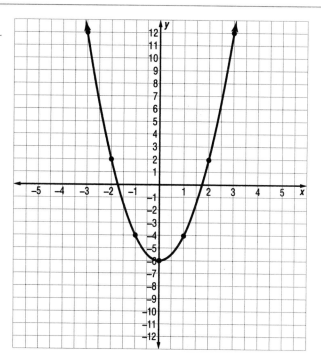

b. approximately $-2\frac{1}{3}$ and $2\frac{1}{3}$

Pages 122–123

1. a–e on graph.

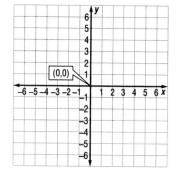

2. Point $A = (-4, 5)$
Point $B = (4, 1)$
Point $C = (2, -4)$
Point $D = (-2, -2)$

3. a.

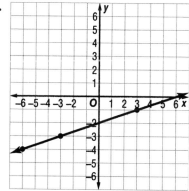

b. x-intercept $= (6, 0)$
y-intercept $= (0, -2)$

c. slope $= \frac{1}{3}$

4. a. 2
b. 5
c. 0

For problem 5, Table of Values will vary.

5.

x	y
0	−4
2	0
4	4

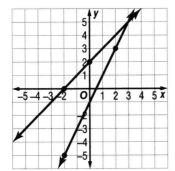

x-intercept $= (2, 0)$
y-intercept $= (0, -4)$
slope $= 2$

6. Common solution $= (3, 5)$

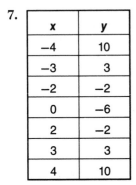

7.

x	y
−4	10
−3	3
−2	−2
0	−6
2	−2
3	3
4	10

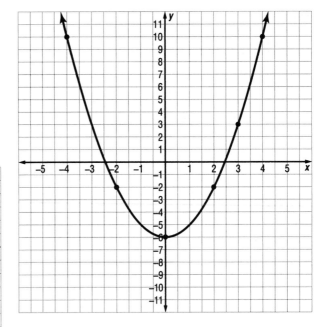

Page 124

1. a. $-3, 0, 5$
b. $-2, 0, 3, 7$

2. a. $8, 10, 19$
b. $6, 11, 23$

3. a. $-8, 4, 12$
b. $-5, 0, 3, 5, 6$

4. a. $17, 29, 41$
b. $-4, -3, 0, 4$

Pages 125–126

1. c. $s \geq -3$

2. a. $r < 1$

3. $s \geq -2$

4. $d < 4$

5.

6.

7.

8.

Pages 128–129

1. $x \geq 2, n \geq 5, z > 4$

2. $w \leq 2, x < -2, y \leq -2$

3. $n > 12, x \geq 40, c < 4$

4. $x \leq 6, n < -8, y \leq 40$

5. $x \leq -8, n > -6, x < -2$

6.
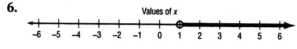
Values of x

7.

Values of n

8
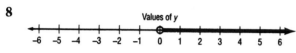
Values of y

9.
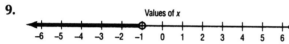
Values of x

10.
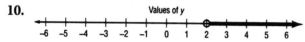
Values of y

Page 130

1. c. $n + 6 > 10, n > 4$

2. a. $3m - 8 \leq 4, m \leq 4$

3. a. $\$1.50n + \$75.00 \geq \$125.00$ or
   $\$1.50n \geq \$50.00$
   b. 34 posters $(n \geq 33\frac{1}{3})$

4. a. $c + 26 + 17 > 75$ or $c + 43 > 75$
   b. 33 yards (The smallest possible value of c is actually a little over 32 yards, so that rounds up to 33 yards.)

5. a. $54w \leq 1,800$
   b. 33 feet

Pages 131–132

1. n is greater than -5 and n is less than 8.

2. x is greater than or equal to -1 and x is less than 11.

3. n is less than or equal to 0 or n is greater than 7.

4. p is less than -3 or p is greater than or equal to 5.

5. a. Yes.
   b. No.
   c. Yes.

6. a. No.
   b. No.
   c. Yes.

7. a. No.
   b. Yes.
   c. Yes.

8. a. Yes.
   b. Yes.
   c. No.

9. a. Yes.
   b. No.
   c. Yes.

10. a. Yes.
    b. Yes.
    c. Yes.

11. $-5 < x \leq 4$

12. $1 \leq r < 3$

13. $y < -2$ or $y \geq 2$

14. $z \leq -3$ or $z > 0$

Page 133

1.

n

2.

x

3. $172 \leq w \leq 178$

4. $0.05(2,000) \leq n \leq 0.1(2,000)$ or $100 \leq n \leq 200$

5. a. $102 < c + 38 + 25 < 119$ or $39 < c < 56$
   b. 40 inches

6. $t < 1$ or $t > 5$

Pages 134–135

1. x is less than 6.

2. n is greater than 9.

3. y is greater than or equal to -5.

4. s is less than or equal to 4.

5. $p > -3$

6. $r \leq -1$

7. $n < 2$

8. $w \geq -5$

9.
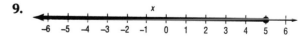
x

10.

n

11. $x \leq 2$

12. $n \geq 10$

13. $n < 20$

14. $x > 8$

15. x is greater than -2 and x is less than 7.

16. y is greater than or equal to 0 and y is less than 6.

17. n is less than or equal to -2 or n is greater than or equal to 5.

18. r is less than 2 or r is greater than 7.

19. $-4 < y \leq 2$

20. $-3 \leq n < 4$

21. **a.** $85 < BC + 31 + 36 < 94$ or $18 < BC < 27$
    b. 19 centimeters

Page 136

1. binomial

2. trinomial

3. monomial

4. binomial

5. trinomial

6. monomial

7. $-12x$ and 7; $9d$ and 6; $-3a^2$ and $5a$; $4y$ and $\frac{1}{2}$

8. $9a$ and $-2b$; $-6d$ and $-8e$; $8x^2$ and $-3y^2$; $\frac{2}{3}z^2$ and $-7z$

9. $2x^2, 7x,$ and -4; $3y^2, -5y,$ and -8; $8x, -4y,$ and 13

Page 137

1. $3x$ and $-5x$; $2y^3$ and $4y^3$; no like terms

2. -5 and -9; $2n$ and $-n$; $4x^3$ and x^3

3. no like terms; $-r$ and $3r$; $-3a^4$ and a^4

4. 36

5. -8

6. 0

7. 5

Pages 138–139

1. $9y, 7r, 19s, 15x$

2. $5z, 8a, 3c, 12n$

3. $5x, 7y, 11r, 0$

4. $-2x^2, 5y^2, 5z^3, 8r^3$

5. $3y + 12, 4x + 13, 7z + 3$

6. $-z + 2, 8y - 8, 16s + 4$

7. $16n^2 + n, 6x^2 + 8x, r^2 + r$

8. $-z + 9, 2x + 11, 4n + 8$

9. $3a - 5b, x + 4y, 5r - s$

10. $3z^2 + 2, 3n^2 + 3, y^2 + 4y$

11. $6a^2 - 2b, 8r^2 - 3s, 13m^2 - 2n$

12. $8m + 2n - 1$

13. $3y^2 + 2x + 1$

14. $4s^2 + 2s - t$

Pages 140–141

1. $3x, 5z, 4r, 7x$

2. $9z, 10b, 8d, 38n$

3. $2x + 2y, 6m + 3n, 5r - 13s$

4. $3x - 8y, 3a + 5b, 12c - 10d$

5. $6a - 4b, 3x - 4y, 7m - 2n$

6. $3x^2 + 4x - 9, 6a^2 + 2a - 9, 7n^2 - 2n - 4$

7. $6a - 5b$

8. $5x^2 - 2x + 7$

9. $3n^2 - n$

Page 142

1. $15y^5, 12x^6, 21n^5, 48a^7$

2. $25x^4, 36n^8, 16z^{12}, 64x^6$

3. $-6x^3, -20c^8, -63b^5, -45s^4$

4. $-7x^3y^2, -5c^4d^4, -12a^4b^5, 56r^3s^4$

Page 143

1. $12n^3 + 8n, 5c^2 + 35c, 12a^4 - 30a$

2. $6x^4 + 12x^3, 8y^3 - 6y^2, 9r^4 + 24r^3$

3. $3a^3b^3 - 6ab, 5c^4d^2 + 3cd, x^3y^4 - 2xy$

4. $-8m^4n^3 - 4m^2n^2, 3a^4b^2 - 18a^3b^2, 15c^3d^3 - 10c^2d^3$

5. $12x^4 - 9x^2y + 3x^2y^2, 16a^3 + 24a^2b^2 - 28ab^3,$
   $6c^4d^4 - 12c^3d^3 + 6c^2d^2$

Page 144–145

1. $8y^2 + 16y + 6, 8a^2 + 8a + 2, 15z^2 + 19z + 6$

2. $8x^2 + 2xy - 3y^2, 6a^2 + 5ab - 6b^2, 3r^2 - 10rs - 8s^2$

3. $a^2 - 4, c^2 - 25, y^2 - 64$

4. $36 - x^2, 81 - b^2, 16 - m^2$

5. $10y^2$

6. $3a^3b - 12ab^2$

7. $c^2 - 36$

8. $64y^3$

9. $x^4 + 3x^2$

10. $9\pi b^3 c^2$

Page 146

1. y^4, x^3, a, c

2. $3b^3, -3n, -4x^3, -7d^4$

3. $a^2b, m^3n^2, -4c^2d, -3x^3y^3$

Page 147

1. $3x^3 + 2x^2, 4n^3 + 2n^2, 5y^4 - 3y^2$

2. $-6n^4 + 2n^3, 4x^4 - 2x^2, -2y^3 - y$

3. $2ab^4 - ab, 3m^2 + 2mn, 4cd - 3c$

4. $6ab^2 - 3b^3 + 4, 3m^2n^3 + 2mn - n^2, 3c^2d - cd + 2d$

Pages 148–149

1. **a.** $n + 4 + n - 2 = 2n + 2$
   b. 26

2. $4r^2 - 2r + 16$

3. **a.** \$4.50n + \$3,500
   b. \$10.50n − \$3,500
   c. \$7,000

4. **a.** $25,000p - p^3$
   b. \$1,125,000

5. **a.** $x^2 + 4x$
   b. 96 square feet

6. **a.** $8x + 48$
   b. $4x^2 + 48x + 144$
   c. 225 square inches

7. **a.** $4x^2 - 60x + 200$
   b. $4x^3 - 60x^2 + 200x$
   c. 144 cubic inches
   d. 192 cubic inches
   e. 168 cubic inches

8. $500 - p$

9. **a.** $20s^2 - 120s - 740$
   b. \$60
   c. $s^2 - 6s - 37$

10. **a.** $n^2 + 4n - 2$
    b. 30

Pages 150–151

1. binomial

2. trinomial

3. monomial

4. binomial

5. $2x$ and $-5x$

6. n^3 and $-2n^3$

7. cd^2 and $-cd^2$

8. $-2a^2b$ and $3a^2b$

9. $11r$

10. $2y^2$

11. $10x - 4$

12. $4n^2 - 2n$

13. $8a + 2b$

14. $3x^2 + 9x - 6$

15. $4s$

16. $17d$

17. $3a + 2b$

18. $6x - 10y$

19. $6m - 3n$

20. $x^2 + x + 5$

21. $20c^8$

22. $6a^3b^4$

23. $8x^4 + 6x$

24. $8m^3 + 2m^2n^2 - 6mn^3$

25. $6a^2 - 11a - 10$

26. $x^2 - x - 20$

27. $35x^2 + 30x$

28. $3a^3b - 6ab^2$

29. n^2

30. $-4a^4b^3$

31. $5c^2 + 2c$

32. $7mn - 4m + 3n$

Page 152

1. 3: 1, 3
   9: 1, 3, 9
   14: 1, 2, 7, 14

2. 24: 1, 2, 3, 4, 6, 8, 12, 24
   36: 1, 2, 3, 4, 6, 9, 12, 18, 36
   64: 1, 2, 4, 8, 16, 32, 64

3. 5, 2, 4, 5

4. 5, 7, 15, 7

5. 2, 2; 3, 3; 5, 2

6. 3, 2; 7, 2; 5, 2

Page 153

1. 1, 11; prime

2. 1, 2, 5, 10; composite

3. 1, 19; prime

4. 1, 3, 7, 21; composite

5. 1, 2, 11, 22; composite

6. 1, 29; prime

7. $2 \times 2 \times 2 \times 2$ or 2^4

8. $5 \times 2 \times 2$ or $2^2 \cdot 5$

9. $2 \times 2 \times 2 \times 2 \times 2$ or 2^5

10. $7 \times 2 \times 2 \times 2$ or $2^3 \cdot 7$

11. $5 \times 5 \times 3$ or $3 \cdot 5^2$

12. $3 \times 2 \times 2 \times 2 \times 2 \times 2$ or $2^5 \cdot 3$

Page 154

1. y, n^2, y^4, r^3

2. $5x^2, 4y^4, 6n^3, 7a^5$

3. $2x^2y^3, 5a^4b, 9c^3d^2, 8x^3y$

Page 155

1. $4\sqrt{2}, 3\sqrt{5}, 5\sqrt{3}$

2. $x\sqrt{y}, x^3\sqrt{y}, b^4\sqrt{a}$

3. $2y\sqrt{x}, 4z^3\sqrt{y}, 5r^2\sqrt{s}$

Pages 156–157

1. $2(x + 3), 3(x + 4), 5(y - 4)$

2. $3(x^2 + 2x + 1), 4(x^2 - 2x + 3), 7(y^2 - 2y - 5)$

3. $x(2x + 5), x(3x - 7), z(2z + 11)$

4. $y(y^2 - 2y + 4), z(z^2 + 6z - 4), x(x^2 - 3x + 2)$

5. $2(w^2 + 4), 3(3x^2 + 1), 4(3y^2 - 2)$

6. $5x(x + 2), 6x(x + 4), 7z(z - 3)$

7. $6z(2z - 3), 2y(2y + 3), 5z(2z - 3)$

8. $3y^2(y + 5), 5x^2(x - 5), 4s^2(s + 4)$

9. $4x^2(2x^2 + 3), 3y^2(3y^2 + 2), 5z^2(3z^2 + 4)$

10. $2ab(b + 2a), 3xy(2x + 1), 2rs(2r + 3s)$

Pages 158–159

1. $x^2 + 2, 4r^2 + 3$

2. $2y + 1, n + 3$

3. $x - 2, 2 - ab$

4. 3, 4

5. x, n

6. $3r, 7v$

7. length $= 4x$

8. width $= 2 + b$

9. height $= 2z^2 + z + 4$

10. base $= 8ab$

Pages 160–161

1. $(5 + s)(5 - s), (6 + t)(6 - t)$

2. $(8 + n)(8 - n), (4 + x)(4 - x)$

3. $(r + 7)(r - 7), (y + 9)(y - 9)$

4. $x(x + 5)(x - 5)$
   $y(y + 6)(y - 6)$
   $n(n + 4)(n - 4)$

5. $4s(s + 3)(s - 3)$
   $8x(x + 2)(x - 2)$
   $2y(y + 4)(y - 4)$

6. $d + 9, t - 8$

7. $5 - r, 4n(n + 3)$ or $4n^2 + 12n$

8. length $= 7 + x$

9. width $= a - 8$

10. height $= 4n(n - 2)$ or $4n^2 - 8n$

Pages 162–163

1. 1, 3, 9
2. 1, 7
3. 1, 2, 3, 4, 6, 8, 12, 24
4. 4, 2
5. 3, 2
6. 7, 3
7. 7×2
8. $2 \times 2 \times 2 \times 2$ or 2^4
9. $5 \times 5 \times 2$ or $2 \cdot 5^2$
10. $4x^2$
11. $6ab$
12. $5c^2d^3$
13. $3\sqrt{2}$
14. $5x\sqrt{y}$
15. $7d^3\sqrt{c}$
16. $4(x + 1)$
17. $3(y - 5)$
18. $6(y + 4)$
19. $x(3x + 7)$
20. $n(2n^2 - 9)$
21. $y^2(5y^2 + 2)$
22. $3x(x + 4)$
23. $5n^2(n - 3)$
24. $2xy(4x + 1)$
25. $x + 1$
26. $n^2 + 2$
27. $2 - mn$
28. 4
29. y
30. $3b$
31. $(6 + n)(6 - n)$
32. $(x + 5)(x - 5)$
33. $x - 7$
34. $3(n + 3)$
35. length $= 3ab + b$
36. width $= x - 6$

16. 25°C
17. **a.** $n - 4 = 10$
18. **a.** $n = -6$
    b. $x = 37$
19. **a.** $p = 9$
    b. $r = 66$
20. **a.** $x = 9$
    b. $a = 15$
21. **a.** $n = 9$
    b. $x = 6$
22. **a.** $s = 13$
    b. $y = 12$
23. length = 57 yards, width = 38 yards
24. 144
25. $x = \pm 6$
26.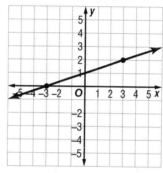

27. slope $= -\dfrac{1}{5}$

28.
x	y
0	−4
2	2
3	5

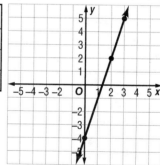

29. $n \le 1$
30. $x \ge 6$
31. $r < 7$
32.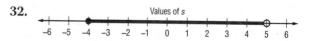

33. $2x^3$ and $-3x^3$

Pages 164–170, Posttest A

1. $-2 + -4 = -6$
2. $5 + (-4) = 1$
3. 15
4. 3
5. −32
6. $\dfrac{5}{9}$
7. 81
8. $\dfrac{1}{64}$
9. 7^6
10. 10^4
11. ± 9
12. $\pm \dfrac{6}{5}$
13. $\dfrac{1}{4}(x - 6)$
14. 14
15. −12

34. $n^2 - 2$

35. $4x^2 - 9x$

36. $8d^3 - 4d^2 + 4d$

37. $7n^3 + 5n$

38. $\pm 6\sqrt{2}$

39. $\pm 7a^3\sqrt{b}$

40. $3(3x^2 + 4)$

41. $5n^2(4n + 3)$

42. $2x^2 + 1$

43. 9

Page 173

1. $-53°$F

2. $-22°$F

3. $-29°$F

4. $6°$F

5. difference $= 27°$

6. difference $= 20°$

7. 35 miles per hour

8. $-5°$F

Page 175

1. 600 miles

2. 55 miles per hour

3. 7 hours

4. 288 miles

5. 52 miles per hour

6. 6 hours

Page 177

1. $30

2. $721.50

3. $95

4. $1,330

5. $71.88

6. $453.75

Page 179

1. $67.50

2. $1,320

3. $85

4. $1,120

5. $112.50

6. $8,102.50

Page 181

1. $\frac{1}{2}$ hour $= 30$ minutes

2. $1\frac{1}{3}$ hours $= 1$ hr 20 min

3. $1\frac{5}{7}$ hours ≈ 1 hr 43 min

4. $2\frac{2}{9}$ hours ≈ 2 hr 14 min

5. $1\frac{7}{8}$ hours ≈ 1 hr 53 min

Page 183

1. 10 pounds of nuts

2. 4 pounds of dried fruit mix

3. 1 pound of the $2.00 per pound chocolates

Pages 184–185

1. $32°$F

2. $77°$F

3. $356°$F

4. $100.4°$F

5. $68°$F

6.

°C	°F
0	32
10	50
30	86

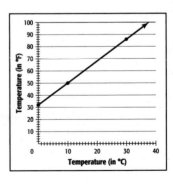

7. about $80°$F (exactly $82.4°$F)

8. about $37°$C (exactly $37.7°$C)

9. about 2 (exactly $\frac{9}{5}$)

Page 187

1. $m\angle EBC = 10°$; $m\angle DBE = 30°$; $m\angle ABD = 50°$

2. $m\angle D = 30°$; $m\angle E = 90°$; $m\angle F = 60°$

3. $m\angle EBC = 6°$; $m\angle DBE = 30°$; $m\angle ABD = 54°$

4. $m\angle A = 18°$; $m\angle B = 72°$; $m\angle C = 90°$

Page 189

1. 10 feet

2. 9 inches

3. 15 feet

4. 13 yards

5. about 8.6 feet

6. about 12 miles

Page 191

1. a. $\frac{h}{250}$
   b. 0.58
   c. 145 feet

2. a. sine $60° = 0.87$
   b. $d = 2.088$ miles

3. a. cosine $45° = 0.71$
   b. 142 miles

Page 193

1. 5 units

2. 10 units

3. 5 units

4. 12 units

5. 13 units

6. $6\frac{1}{2}$ units

Page 195

1. $4, 0, 8.5, \frac{1}{2}, 9$

2. $2, 9, 7$

3. $x = -5$ and $x = 5$
   $n = -3$ and $n = 3$
   $r = -10$ and $r = 10$
   $w = 0$
   $z = -100$ and $z = 100$

4. $x = -1$ and $x = 5$
   $n = 4$ and $n = 6$
   $r = -8$ and $r = 22$

5.

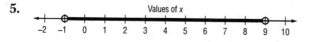
Values of x

6.

Values of n

7. $|r - 2| < 5$

Pages 196–201, Posttest B

1. e. $4 + (-3) = +1$

2. c. $-24°F$

3. b. 16

4. e. 16

5. a. 6^5

6. b. $\pm\frac{6}{5}$

7. c. $4(n - 7)$

8. d. 40

9. a. $10°C$

10. c. $n - 8 = 19$

11. e. Add $+8$ to each side of the equation.

12. d. Subtract 9 from each side of the equation.

13. a. $x = 18$

14. c. $n = 4$

15. d. $x + (x + \$40) + 2x = \$2,620$

16. c. 55

17. c. two

18. d. $\frac{n}{18} = \frac{3}{5}$

19. e. $(4, 0)$

20. b. -1

21. c. $(1, 2)$

22. e. 6

23. d. $x < 1$

24. b. $-4 < x \le 4$

25. a. $3n^2 - 3n$

26. e. $13x^2 + 8$

27. d. $x^4 + 4x^3$

28. a. $7x^3\sqrt{y}$

29. d. $8n^2(2n^2 - 3)$

30. e. $r - 3$

FORMULAS

PERIMETER

Figure	Name	Formula	Meaning
	Rectangle	$P = 2l + 2w$	l = length w = width
	Square	$P = 4s$	s = side
	Triangle	$P = s_1 + s_2 + s_3$	s_1 = side 1 s_2 = side 2 s_3 = side 3
	Polygon (n sides)	$P = s_1 + s_2 + \ldots + s_n$	s_1 = side 1, and so on
	Circle (circumference)	$C = \pi d$ or $C = 2\pi r$	$\pi \approx 3.14$ or $\frac{22}{7}$ d = diameter r = radius

AREA

Figure	Name	Formula	Meaning
	Rectangle	$A = lw$	l = length w = width
	Square	$A = s^2$	s = side
	Parallelogram	$A = bh$	b = base h = height
	Triangle	$A = \frac{1}{2}bh$	b = base h = height
	Circle	$A = \pi r^2$	$\pi \approx 3.14$ or $\frac{22}{7}$ r = radius

VOLUME

Figure	Name	Formula	Meaning
	Rectangular solid	$V = lwh$	l = length w = width h = height
	Cube	$V = s^3$	s = side
	Cylinder	$V = \pi r^2 h$	$\pi \approx 3.14$ or $\frac{22}{7}$ r = radius h = height

OTHER FORMULAS

Name	Formula or Example	Meaning
Difference of squares	$a^2 - b^2 = (a + b)(a - b)$	factors of a difference of squares
Distance formula	$d = rt$	distance = rate $\times$ time
Dividing like bases	$\dfrac{x^a}{x^b} = x^{a-b}$	dividing like bases
Simple interest	$i = prt$	i = interest, p = principal, r = rate (%), t = time (yr)
Linear equation	$y = 4x - 1$	two-variable equation in which each variable has an exponent of 1; m and b are numbers
Multiplying like bases	$x^a x^b = x^{a+b}$	multiplying like bases
Pythagorean theorem	 $c^2 = a^2 + b^2$	c = hypotenuse; a and b are the two shorter sides of the right triangle
Quadratic equation	$y = 2x^2 + 3$	two-variable equation in which one variable is squared
Slope	$\dfrac{\text{change in } y \text{ values}}{\text{change in } x \text{ values}}$	slope of a line on a coordinate grid
Temperature	$°F = \dfrac{9}{5} °C + 32$	$°F$ = degrees Fahrenheit $°C$ = degrees Celsius

GLOSSARY

A

absolute value   The positive value of a number or expression regardless of its sign. For example, $|5| = 5$ and $|-5| = 5$

algebraic expressions   Two or more numbers or variables combined by addition, subtraction, multiplication, or division

***and* inequality**   An inequality, such as $2 \leq x < 6$, in which the allowed values of the variable run between two numbers

area (A)   An amount of surface taken up by a two-dimensional object. Area is measured in square units.

axis   One of the perpendicular sides of a graph along which numbers, data values, or labels are written. Plural is *axes*.

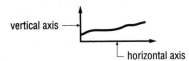

B

base   The number being multiplied in a power. In 2^3, 2 is the base; 3 is the exponent.

binomial   A polynomial of two terms. Examples: $x + 4, 3y^2 - 1, r^3 + 5$

C

circle   A plane (2-dimensional) figure, in which each point is an equal distance from the center

circumference   The distance around a circle. $C = \pi \times$ diameter

coefficient   A number that multiplies a variable. In the term $3y$, 3 is the coefficient.

comparison symbols   Symbols used to compare one number with another

Symbol	Meaning	Example	Read As
$<$	is less than	$n < 7$	n is less than 7
$>$	is greater than	$x > 6$	x is greater than 6
$\leq$	is less than or equal to	$y \leq 2$	y is less than or equal to 2
$\geq$	is greater than or equal to	$s \geq -3$	s is greater than or equal to -3

composite number   A number, such as 12, that has more than two factors

coordinates   A set of two numbers in the form (x, y) that identifies a point on a coordinate grid

coordinate axis   A number line that forms the horizontal axis or vertical axis of a coordinate grid. See *horizontal axis* and *vertical axis*.

coordinate grid   A grid on which points are identified by an ordered pair of numbers (x, y), called coordinates

cosine of an angle   In a right triangle, the ratio of the near side of an angle to the hypotenuse.

cosine $\angle A =$ near side/hypotenuse

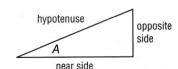

cross products   In a proportion, the product of the numerator of one ratio times the denominator of the other. If the proportion is true, cross products are equal.

Proportion	**Equal Cross Products**
$\dfrac{2}{3} = \dfrac{6}{9}$	$2 \times 9 = 3 \times 6$
	$18 = 18$

cube   A 3-dimensional shape that contains six square faces. At each vertex, all sides meet at right angles.

cube

cylinder   A 3-dimensional shape that has both a circular base and a circular top

cylinder

D

degree   The measure (size) of an angle. A circle contains 360 degrees (360°). One-fourth of a circle contains 90°.

diagonal   A line segment running between two nonconsecutive vertices of a polygon. A diagonal divides a square or rectangle into two equal right triangles.

diagonal

diameter   A line segment, passing through the center, from one side of a circle to the other. The length of the diameter is the distance across the circle.

diameter

distance formula   The formula $d = rt$ relates distance, rate, and time. In words, *d*istance equals *r*ate times *t*ime. By rearranging the variables, you can write the rate formula $r = \frac{d}{t}$ or the time formula $t = \frac{d}{r}$.

E

equation   A statement that two quantities are equal (have equal value or measure)

equivalent expressions   Expressions that have the same value. The expression $2x + 4x + 5$ is equivalent to the expression $6x + 9 - 4$. Both expressions simplify to $6x + 5$.

exponent   A number that tells how many times the base (of a power) is written in the product. For example, in 5^2, 2 is the exponent and means that there are two 5s in the product, 5×5.

estimate   To find an approximate answer. For example, by calculating with rounded numbers.

F

factor   A number that divides evenly into another number. Numbers and variables can be factors of an algebraic expression. Example: 1, 2, 4, and 8 are factors of 8.

formula   A mathematical rule that tells the relationship between quantities. For example, the formula relating the area of a circle to its radius is $A = \pi r^2$.

H

horizontal   Running right and left

horizontal axis   On a coordinate grid, the axis running from left to right

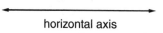

horizontal axis

hypotenuse   In a right triangle, the side opposite the right angle

hypotenuse

I

inequality   A statement that two quantities are not equal. Examples: $n \geq 6$ states that n is greater than or equal to 6; $-2 < x \leq 3$ states that x is greater than -2 and less than or equal to 3.

intercept   A point where a graphed line crosses a coordinate axis. The x-intercept is the point where the line crosses the x-axis. The y-intercept is the point where the line crosses the y-axis.

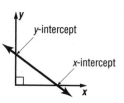

inverse operation   An operation that "undoes" another operation: subtraction is the inverse operation of addition; division is the inverse operation of multiplication.

L

like terms   Terms that have the same variable raised to the same power: 4 and 7 are like terms (no variable); $2r$ and $5r$ are like terms; $3x^2$ and $-2x^2$ are like terms.

linear equation   An equation such as $y = 2x$ that, when graphed, forms a straight line

M

monomial   A polynomial of one term. Examples: $x, y^3, 5r^2$

multistep equation   An equation such as $2x + 5 = 9$ that contains more than one operation

N

negative exponent   An exponent that is less than 0. A negative exponent stands for the number found by inverting a power.

negative number   A number less than zero. A negative number is written with a negative sign. Examples: $-6, -2.5, -1$

number line   A line used to represent positive numbers, negative numbers, and 0. A number line can be written horizontally (left to right) or vertically (up and down).

O

***or* inequality**   An inequality, such as $r < 4$ *or* $r \geq 9$, in which the allowed values of the variable are graphed as separate sections of a number line

ordered pair   A pair of numbers of the form (x, y) used to identify a point on a coordinate grid

P

perfect square   A number whose square root is a whole number. Example: 64 is a perfect square; $\sqrt{64} = \pm 8$.

perimeter   The distance around a flat (2-dimensional) figure. The symbol for perimeter is P.

perpendicular lines   Lines that meet (or cross) at a right angle (90°)

perpendicular line

pi (π)   The ratio of the circumference of a circle to its diameter. Pi is approximately equal to 3.14 or $\frac{22}{7}$.

polynomial   An algebraic expression that contains one or more terms. Monomials, binomials, and trinomials are all polynomials.

positive number   A number greater than 0. Positive numbers may be written with a positive sign ($+$) or no sign at all.

power   The product of a number multiplied by itself one or more times. Example: $5^2 = 5 \times 5 = 25$

prime factorization form   Writing a number as a product of prime factors. Example: $36 = 2 \times 2 \times 3 \times 3 = 2^2 \times 3^2$

prime number   Any number greater than 1, such as 7, that has only two factors: itself and the number 1

proportion   Two equal ratios. A proportion can be written with colons or as equal fractions.

Written with colons: $2:3 = 6:9$

Written as equal fractions: $\frac{2}{3} = \frac{6}{9}$

Pythagorean theorem   A theorem that states: In a right triangle, the square of the hypotenuse is equal to the sum of the squares of the two shorter sides. Also stated as $c^2 = a^2 + b^2$.

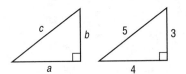

Q

quadratic equation   An equation such as $y = x^2$ that contains the square of a variable

R

radius   A line segment from the center of a circle to any point on the circumference of the circle

radius

ratio   A comparison of two numbers. For example, "4 to 3." A ratio can be written as a fraction or with a colon.

4 to 3 is written $\frac{4}{3}$ or 4:3

reciprocal   A number formed by inverting a fraction or whole number: $\frac{3}{2}$ is the reciprocal of $\frac{2}{3}$; $\frac{1}{4}$ is the reciprocal of 4.

rectangle   A four-sided polygon with two pairs of parallel sides and four right angles

rectangle

rectangular solid (prism)   A 3-dimensional figure in which each face is either a rectangle or a square. Opposite faces are congruent.

rectangular solid

reflex angle   An angle that measures greater than 180° but less than 360°

reflex angle

right angle   An angle that measures exactly 90°. A right angle is often called a "corner angle."

right angle

right triangle   A triangle that contains a right angle

right triangle

S

scientific notation   Writing a number as a product of two factors: a number between 1 and 10 *and* a power of 10. Examples: $5{,}300 = 5.3 \times 10^3$, $0.075 = 7.5 \times 10^{-2}$

signed numbers   The set of all negative numbers and positive numbers

simple interest   Interest earned (or paid) on an unchanging principal. The simple interest formula is $i = prt$.

sine of an angle   In a right triangle, the ratio of the opposite side of an angle to the hypotenuse.

sine $\angle A$ = opposite side/hypotenuse

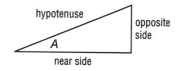

hypotenuse   opposite side
A
near side

slope   The rise or fall of a graphed line. The slope is defined as the change in y values divided by the change in x values.

Positive slope: The line rises (goes up) from left to right.

Negative slope: The line falls (goes down) from left to right.

Zero slope: A horizontal line that neither rises nor falls

Undefined slope: A vertical line

square   A polygon with four equal sides, two pairs of parallel sides, and four right angles

square

square of a number   The product of a number multiplied by itself. Example: $8^2 = 8 \times 8 = 64$

square root ($\sqrt{\ }$)   One of two equal factors of a number. Example: $\sqrt{9} = \pm 3$

system of equations   Two or more linear equations. Example: $y = 2x + 3$ and $y = x + 2$ have the common solution $x = -1$ and $y = 1$.

T

tangent of an angle   In a right triangle, the ratio of the opposite side of the angle to the near side

tangent $\angle A$ = opposite side/near side

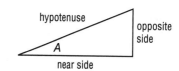

hypotenuse   opposite side
A
near side

term   In a polynomial, a number standing alone or a variable multiplied by a coefficient. Examples: 5, $3x$, $2y^2$. Terms are connected by $+$ and $-$ signs.

triangle   A polygon having three sides and three angles

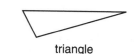

triangle

trigonometric function   Referring to a particular angle, the ratio of any two sides of a right triangle. See *cosine of an angle, sine of an angle,* and *tangent of an angle.*

trigonometry   The study of the relations of the sides and angles of a right triangle

trinomial   A polynomial of three terms. Examples: $x^2 - 3x + 1, 2y^3 + y + 4$

U

unknown   A letter that stands for a number. Also called a *variable.* For example, in $x + 4 = 7$, x is an unknown standing for the number 3.

V

variable   A letter that stands for a number. Also called an *unknown*.

vertical axis   On a coordinate grid, the axis running up and down

vertical axis

volume   The amount of space an object takes up. Volume (V) is measured in cubic units.

W

whole numbers   The set of counting numbers (1, 2, 3, ...) and the number 0

windchill   A temperature reading that takes into account the chilling effect of the wind. Windchill temperature is lower than thermometer temperature.

work problem   A problem that involves finding the rate at which two or more people working together can finish a job

INDEX

P

Q

R

S

T

U

V

W